Alfred James McClatchie

Flora of Pasadena and vicinity

Alfred James McClatchie

Flora of Pasadena and vicinity

ISBN/EAN: 9783337268503

Printed in Europe, USA, Canada, Australia, Japan

Cover: Foto ©berggeist007 / pixelio.de

More available books at **www.hansebooks.com**

FLORA

OF PASADENA

AND VICINITY . . . .

By ALFRED JAMES McCLATCHIE, A. M.,
Professor of Botany,
Throop Polytechnic Institute.

[REPRINTED FROM DR. H. A. REID'S HISTORY OF PASADENA.]
Issued in 1895

PRICE, 25 CENTS.

1895
BARNES & NEUNER CO. PRINT
LOS ANGELES, CAL.

By Alfred James McClatchie, A. M.,

Professor of Botany, Throop Polytechnic Institute.

The list of plants that follows was compiled, not because it was supposed that all of the plants growing about Pasadena were known, but because of the kind and urgent request made by Dr. Reid that I should undertake the task. Our flora is too varied for one person to become familiar with it during a three years' residence. The list simply includes all that have been collected and identified up to the time of going to press ; but each month adds several to the list, and will, undoubtedly, continue to do so for some time yet.

The region of which the plants are listed extends from the Lincoln Park hills on the south to the summit of the range north of Pasadena, designated by various names, but referred to in this list as the San Gabriel mountains. On the west the region is bounded by the hills across the Arroyo Seco and extends from there eastward to Sierra Madre and Santa Anita. Thus it is about ten miles in extent north and south, and about six miles east and west. The altitude at the southern limit is about 500 feet, while the summit of the mountains varies from 5,000 to 6,000 feet. The altitudes at the eastern and western boundaries are about the same. Hence the region might be thought of in a general way as a surface having a parabolic curve, one end resting against a range of low hills and the other resting upon a support ten miles away and a mile higher. The western edge of the region is traversed by the Arroyo Seco, whose precipitous banks average about fifty feet in height. At the bottom flows a swift stream, notwithstanding the fact that the name is the Spanish for "dry gorge." About a mile south of Pasadena is the lip of the geological basin that has been filled with soil for a site for our city. From this lip flow six nearly parallel streams, each about one-fourth to one-half mile from the next. Along these streams grow luxuriant forests of oak, sycamore, cottonwood, and alder, among which great numbers of higher fungi flourish during the wet season, and their waters abound in algæ and other water plants. To this region Dr. Reid and Mr. French have given the name Glacial Terrace. [See page 574.]

Between this lip and the foot of the mountains extends a sloping plain which bore, when in its natural state, principally herbaceous plants and small shrubs, some of which still remain scattered over the region. The mountain region is traversed by two large canyons that begin near the

summit of Mount Lowe, besides several smaller ones that do not begin so
far back in the mountains. It will readily be seen that a region so diversi-
fied — a region made up of mountain, canyon, plain, and moist woodland —
must have a varied flora. In the woods and canyons algæ, fungi, liverworts, mosses, ferns, and shade-loving seed-plants abound; on the plain,
dwarfed shrubs, cacti, and other plants characteristic of an arid region; on
the mountains, large shrubs, live-oaks, pines, spruces, cedars, and a great
variety of smaller plants.

As an examination of the list will show, every month of the year brings
forth some new plants, and during every month some of the higher plants
are in bloom. The season of greatest growth is from January to June. It
is during this period that the most of the lichens, the higher fungi, the
liverworts, the mosses, the ferns, and the herbaceous seed-plants grow and
reproduce. During the remainder of the year parasitic fungi flourish, a
few mosses mature their spores, several Polygonaceæ, Euphorbiaceæ, Chenopodiaceæ, Compositæ, and some members of other families of seed-plants
bloom and mature seeds. Algæ are to be found at all seasons of the year
where there is sufficient water. Two beautiful species of Florideæ grow in
abundance throughout the year in the Arroyo Seco, and a great variety of
brook-silk, green-felt, water-flannel, and other green algæ is always to be
found there. The lover and student of plants can find an abundance of
specimens to enjoy and study at all seasons of the year.

The plants of California, especially of the southern part, have not yet
been so carefully listed and described as in many of the eastern and
southern states. The literature accessible for their identification is still
meager. The Botany of the Geological Survey, the works of Professor
E. L. Greene of the State University, and some works descriptive of all the
plants of certain groups found in North America, are the more useful books.
The following are the principal works that list or describe plants of our
region: Sternberg's Manual of Bacteriology, Wolle's Fresh-water Algæ,
Wolle's Desmidiaceæ, Wolle's Diatomaceæ, Pound's Mucoreæ of N. A.,
Ellis and Everhart's Pyrenomycetes, Tuckermann's Lichens, Saccardo's
Sylloge Fungorum, Kellerman & Ellis's Journal of Mycology, Morgan's
Gastromycetes, Farlow and Seymour's Host-Index, Harvey's Nereis
Borealis, Hervey's Sea-mosses, Halsted's Characeæ of America, Allen's
Characeæ of North America, Underwood's Hepaticæ of North America,
Lesquereux and James's Mosses of N. A., Eaton's Ferns of N. A., Under-
wood's Our Native Ferns and Their Allies, Jones's Ferns of the Pacific
Coast, Vasey's Grasses of the Southwest, Bailey's Carices of N. A.,
Morong's Naiadaceæ of N. A., Watson's Liliaceæ of N. A., Greene's Oaks
of the Pacific Coast, Wheelock's Polygala, Trealease's Rumex, Trealease's
Epilobium, Greene's Pittonia, Brewer, Watson and Gray's Botany of Cali-
fornia, Gray's Synoptical Flora of N. A., Rattan's Popular Flora, Parish's
Plants of Southern California, Davidson's Plants of Los Angeles county.

Besides the above, several works describing plants of other regions are useful. Among them are Burrill & Earle's Parasitic Fungi of Illinois, Rabenhorst's Cryptogamic Flora of Germany, Austria and Switzerland, Massee's British Fungus-Flora, Cooke's British Fungi, Berkeley's Outlines of Mycology, Phillip's Discomycetes, Farlow's Marine Algae of New England, Greene's Flora Franciscana, Greene's Manual of the Bay Region, Gray's Manual of Botany, Coulter's Rocky Mountain Botany, Chapman's Flora of the Southern States, Coulter's Flora of Texas, Wood's Class-book of Botany, and the Flora of Nebraska by the Botanical Seminar of the State University.

Several plant catalogues of other regions aid much in classification and nomenclature. Among these are Britton's Flora of New Jersey, Wheeler & Smith's Flora of Michigan, Webber's Flora of Nebraska, McMillan's Metaspermæ of Minnesota, Millspaugh's Flora of West Virginia, and the Pteridophyta and Spermophyta of the Northeastern United States by the Botanical Club of the Am. Assoc. for Advct. of Sci.

All plants of doubtful identity have have been sent to specialists for determination, as follows: Perisporiaceæ and Pyrenomycetes to J. H. Ellis, Newfield, N. J.; Discomycetes to A. P. Morgan, Preston, Ohio. J. B. Ellis and C. H. Peck, Albany, N. Y.; Lichenes to T. A. Williams, Brooking, S. D.; Uredineæ to E. W. D. Holway, Decorah, Iowa; Imperfect Fungi to J. B. Ellis; Gastromycetes to A. P. Morgan and L. M. Underwood, Greencastle, Ind.; Agaricineæ to C. H. Peck and F. E. Clements, Lincoln, Neb.; the remaining Hymenomycetes to A. P. Morgan, L. M. Underwood, and J. B. Ellis; Hepaticæ to L. M. Underwood; Musci to C. R. Barnes, Madison, Wis., Mrs. E. G. Britton, Columbia College, N. Y., and M. A. Howe, Berkeley, Cal.; Pteridophyta to L. M. Underwood, and D. C. Eaton, New Haven, Conn.; Gramineæ to F. Lamson-Scribner, Washington, D C.; Carices to L. H. Bailey, Ithaca, N. Y.; the remaining Spermaphyta to E. L. Greene, Berkeley, Cal., S. B. Parish, San Bernardino, Cal., and W. L. Jepson, Berkeley, Cal. To all I am greatly indebted for their aid, and wish to express to them my sincere thanks. I have also received much aid from several of my students, especially Miss Dian Haynes and Miss Margaret Morrison. To my wife, Anna Morrison McClatchie, I am especially indebted for continuous aid in collecting, identifying, drawing, and caring for herbarium specimens.

Unless impracticable, herbarium or microscopic specimens of each species listed have been preserved. Duplicates of a large number of them will also be found in the herbaria of those who have aided in their indentification. The aim has been to give the local place where all species not widely distributed have been collected.

The month or months given as the season of a plant cover the period of reproduction, or when reproductive organs may be found on the plant.

When no period is given, it is to be understood that one or both the above conditions are present all of the year. The elevation of many of the plants is given, and for most of the others the elevation can be inferred from the place where found, by any one familiar with the region. I have also aimed to give the common name of each plant having a good one. When the specific name of a parasitic fungus is formed from the generic name of the host-plant, the initial letter has been used for the genus of the latter. All Agarics stated to be edible have been tested by myself, and several of them by some of my students.

Of most of the Bacteria listed, and many others not yet identified, pure cultures have been made in my laboratory. The pathogenic Bacteria are not listed. Little effort has been made to identify the Diatoms of the region, hence few of them are listed. Of the Agarics, about fifty collected species remain undetermined; of Lichens, about ten; and of Mosses, about the same number. The number of species and varieties listed is 1056, of which a large number were never before collected in the State. Sixty-two of them proved to be new to science. Most of these have been described by group specialists and by myself, in the botanical journals and in the proceedings of scientific societies of America. The place of publication of each new species is cited in the list. Being opposed to the naming of new species after collectors, I have attempted to prevent any being given my name, and have succeeded in all cases except one that was published in spite of my protest.

I have attempted to follow the Rochester rules for nomenclature, but no doubt have failed to do so in many cases. The system of classification used is, in the main, that of Dr. Bessey. In the groups below the Spermaphyta, no smaller subdivision than Bessey's orders have been used above genera. In the Spermaphyta, family names are used, Bessey's ordinal names being omitted.

THROOP BIOLOGICAL LABORATORY,
PASADENA, CAL., September 26, 1895.

PROTOPHYTA.

CLASS I. MYCETOZOA. Slime Moulds.

RETICULARIA Bull.
R. UMBRINA Fries. On decaying wood. Frequent. March — May.
HEMIARCYRIA Rost.
H. RUBIFORMIS (Pers.) Rost. On decaying wood. M. C.* January — May.
TRICHIA Hall.
T. VARIA Pers. On decaying wood. M. C. January — April.
STEMONITIS Gled.
S. FUSCA Roth. On decaying wood. Frequent. March — September.

* ABBREVIATIONS.—The following abbreviations have been used to designate particular places by local name, where specimens have been found: Lincoln Park L. P.; Arroyo Seco, A. S.; Los Robles Canyon, L. R. C.; Oak Knoll, O. K.; Oak Knoll Canyon, O. K. C.; Wilson Canyon, W. C.; Wild Grape Canyon [see page 377], W. G. C.; San Gabriel Mountains, S. G. Mts.; Millard Canyon, M. C.; Rubio Canyon, R. C.; Little Santa Anita Canyon, L. St. A. C.; Wilson's Peak, W. Pk.; Mount Lowe, Mt. L.

SPUMARIA Pers.

S. ALBA (Bull.) DC. On living willow stems. A. S. February — October.

DIACHEA Fries.

D. LEUCOPODA (Bull.) Rost. On dead and living leaves. Com. January — April

BADHAMIA Berk.

B. HYALINA Berk. On decaying wood in L. R. C. February — April.

FULIGO Hall.

F. SEPTICA (Link.) Gmel. On decaying wood. Frequent. February — April.

PHYSARUM Pers.

P. CINEREUM (Batsch.) Pers. On small living plants. January — March.

CLASS II. SCHIZOPHYCEÆ. Fission-Plants.

Order Cystiphoræ. One celled blue-green Algae.

CHROOCOCCUS Naeg.

C. COHOERENS (Breb.) Naeg. Common in stagnant water.

MERISMOPEDIA Mey.

M. GLAUCA (Ehrb.) Naeg. In watering-trough.

Order Nematogenæ. Filamentous blue-green Algae, Bacteria, etc.

NOSTOC Vauch.

N. MUSCORUM Ag. Frequent among moss. January — April.
N. PRUFIFORME (Roth.) Ag. Common in running water.
N. RUPESTRE Kuetz. Among moss on moist banks.
N. SPHAERICUM Vauch. On wet soil. January — April.

ANABAENA Bory.

A. STAGNALIS Kuetz. In moist banks. June — October.

OSCILLARIA Bosc.

O. ANTLIARIA Juerg. In reservoir.
O. BREVIS Kuetz. On wet soil.
O. MAJOR Vauch. In pond at Oak Knoll. A. S.
O. TENERRIMA Kuetz. In stagnant water in Arroyo Seco.
O TENUIS Ag. Common. In water and on wet soil.

LEPTOTHRIX Kuetz.

L. CAESPITOSA Kuetz. In watering-trough.

CYLINDROSPERMUM Kuetz

C. COMATUM Wood. In stagnaut water in Arroyo Seco.
O. FLEXUOSUM (Ag.) Rab. Among moss on moist banks.
O. MACROSPERMUM Kuetz. In stagnant water in Arroyo Seco.

LYNGBYA Ag. & Thur.

L. OCHRACEA (Dill) Thur. On a moist bank.

TOLYPOTHRIX Kuetz.

T. DISTORTA (Muell.) Kuetz. Rubio Canyon — on rocks under running water.

MASTIGONEMA (Fisher) Kirch.

M. AERUGINEUM (Kuetz.) Kirch. Among damp moss
M. FERTILE Wood. In reservoir.

HAPALOSIPHON Naeg.

H. BRAUNII Kuetz. In aquarium in laboratory.

BEGGIATOA Trevisan.

B. ALBA (Vauch.) Trev. Frequent in stagnant water.

MICROCOCCUS Cohn.

M. CREPUSCULUM (Ehrb.) Cohn In decaying fish.

ASCOCOCCUS Zopf.

A. BILROTHII Stern. In putrid vegetable infusion.

BACILLUS Cohn.

B. ACETI (Kuetz.) Cohn. In vinegar.
B. ACIDI LACTICI Hueppe. In sour milk.

39

B. FLUORESCENS LIQUEFACIENS Fluegge. Common in water.
B. TERMO (Muell.) Cohn. Common in various decaying substances.
B. VULGARIS Haus. In decaying meat.
SPIRILLUM Ehrb.
   S. RUGULA (Muell.) Ehr. Common in decaying substances.

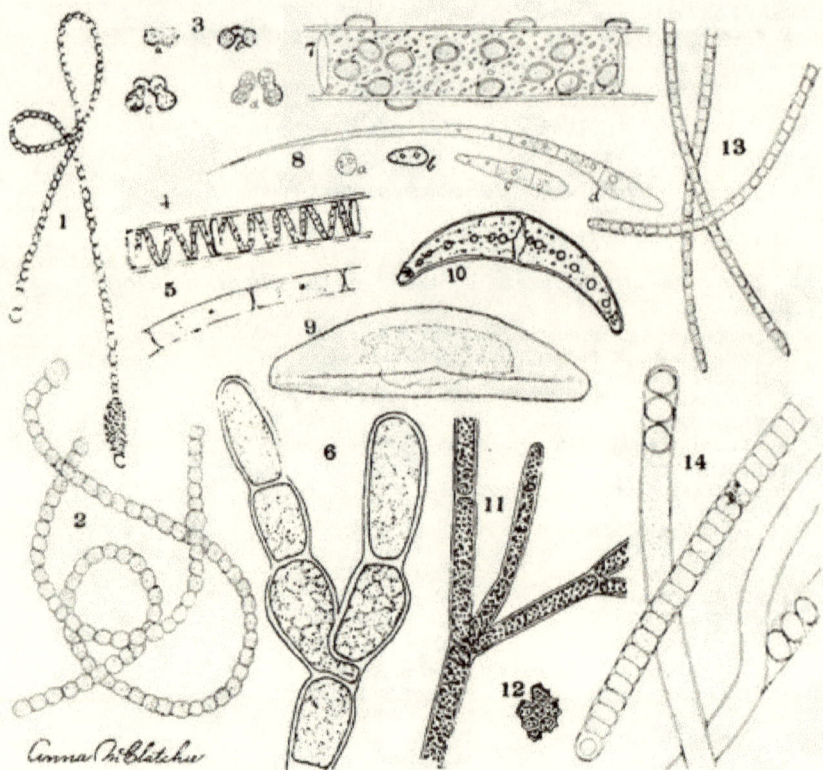

FRESH-WATER ALGÆ.

1. *Cylindrospermum flexuosum.* 2. *Nostoc muscorum.* 3. *Cosmarium cordanum* in process of division. 4. *Spirogyra adnata.* 5. *Zygnema stellium* 6. *Cladophora fracta.* 7. *Coconeis pediculus* ou *Cladophora.* 8. Spores of *Stigeoclonium fastigiatum* in various stages of germination. *Cymbella gastroides.* 10. *Closterium moniliferum.* 11. *Cladophora oligoclona.* 12. *Pediastrum boryanum.* 13. *Oscillaria antillaria.* 14. *Tolypothrix distorta.* All magnified 250 diameters.

PHYCOPHYTA.

CLASS I. CHLOROPHYCEÆ.

Order Protococcoideæ.

PROTOCOCCUS Ag.
   P. VIRIDIS Ag. Very common in water and on wet surfaces.
SCENEDESMUS Meyen.
   S. DIMORPHUS Kuetz. Common in stagnant water.
   S. OBTUSUS Mey. In aquarium in laboratory.
   S. ACUTUS Mey. Frequent in stagnant water.

PEDIASTRUM MEYEN.
P. BORYANUM (Thurp.) Menegh. Frequent in stagnant water.
HYDRODICTYON Roth.
H. UTRICULATUM Roth. Common in streams.
PANDORINA Ehrb.
P. MORUM Bory. Common in stagnant water.
GONIUM Muell.
G. PECTORALE Muell. In stagnant water in Arroyo Seco.
EUDORINA Ehrb.
E. STAGNALE Wolle. In stagnant water in Arroyo Seco.
EUGLENA Ehrb.
E. VIRIDIS (Schrank.) Ehrb. Frequent in stagnant water.

Order. Conjugatæ. Desmids, Diatoms, etc.

CLOSTERIUM Nitsch.
C. MONILIFERUM (Bory.) Ehrb. Frequent in stagnant water.
C. ENSIS Delp. In stagnant water at Oak Knoll.
CALOCYLINDRUS D. By.
C. CONNATUS (Breb.) Kirch. var. MINOR Nord. Frequent in stagnant water.
COSMARIUM Corda.
C. CORDANUM Breb. In stagnant water in Arroyo Seco.
CYMBELLA Agardh.
C. GASTROIDES Kuetz. Common in stagnant water.
COCCONEMA Ehrb.
C. MEXICANUM Ehrb. Common in stagnant water.
NAVICULA Bory.
N. SUBINFLATA Grun. In water in Arroyo Seco.
GOMPHONEMA Agardh.
G. ACUMINATUM Ehrb. In stagnant water in Arroyo Seco.
COCONEIS Ehrb.
C. PEDICULUS Ehrb. On Cladophora in M. C.
C. CALIFORNICA Grun. In stagnant water in Arroyo Seco.
SYNEDRA Ehrb.
S. VALENS Ehrb. Common in stagnant water.
MERIDION Agardh.
M. CIRCULARE (Grev.) Ag. Common in stagnant water.
MELOSIRA Agardh.
M. VARIANS Ag. Common in stagnant water.
MESOCARPUS Hass.
M. RADICANS Kuetz. Stream in Arroyo Seco.
M. SCALARIS (Hass.) D. By. In stagnant water in Arroyo Seco.
ZYGNEMA Kuetz.
Z. STELLIUM Ag. On wet rocks in Rubio canyon.
SPYROGYRA Link. Pond-scum. "Frog-spittle." Brook-silk.
S. AENATA Kuetz. Frequent in stagnant water.
S. FUSCO-ATRA Rab. In stagnant water in Arroyo Seco.
S. ORTHOSPIRA (Naeg.) Kuetz. Frequent in stagnant and running water.
S. CRASSA Kuetz. Frequent in stagnant water.
S. QUININA (Ag.) Kuetz. Com. in stagnant water. Conjugates during April and May.
ASCOPHORA Tode.
A. MUCEDO Tode. (Common black mould.) Common on decaying substances.
MUCOR Linn. Black mould.
M. MUCEDO Linn. Occasional on decaying substances.
M. RACEMOSUS Fres. On decaying cooked onion.
EMPUSA Cohn.
E MUSCAE (Fr.) Cohn. (Fly fungus) On flies.

Order. Siphoniæ.

VAUCHERIA DC. Green-Felt.
  V. HEMATA (Vauch. Lyng. Common in running and stagnant water.
  V. SESSILIS (Vauch.) DC. In pond in W. C.
  V. TERRESTRIS Lyng. Frequent on moist soil.
SAPROLEGNIA Nees. Water-mould.
  S. FERAX (Gruith. Nees. On flies in aquarium in laboratory.
BOTRYDIUM Wallr.
  B. GRANULATUM (L.) Grev. Common on moist soil. January — May.
ALBUGO S. F. Gray. White rust.
  A. CANDIDA (Pers.) OK. On Shepherd's Purse. February — May.

Order — Confervoideæ. Water-flannel, etc.

CLADOPHORA Kuetz. Water-flannel.
  C. FRACTA (Dill) Kuetz. var. RIGIDULA Kuetz. Frequent in stagnant and running
    water.
  C. OLIGOCLONA Kuetz. Common in stagnant and running water.
ULOTHRIX Kuetz.
  U. ZONATA (W. & M.) Ag. In watering-trough.
CONFERVA Link.
  C. FLOCCOSA Ag. Common in stagnant and running water.
  C. FUGACISSIMA Roth. Frequent in stagnant and running water.
STIGEOCLONIUM Kuetz.
  S. FASTIGIATUM Kuetz. Common in stagnant and running water.
CYLINDROCAPSA Rein.
  C. GEMINELLA Wolle. On shaded soil in Pasadena. February — April.
DRAPERNALDIA Agardh.
  D. GLOMERATA (Vauch.) Ag. Frequent in stream in Arroyo Seco. March — May.
ŒDOGONIUM Link.
  Œ. AUTUMNALE Witt. Stagnant water at Oak Knoll.

CARPOPHYTA.

CLASS II. ASCOMYCETES. Sac-Fungi.

Order Perisporiaceæ. Simple Sac-Fungi.

SPHÆROTHECA Lev.
  S. PANNOSA (Wallr.) Lev. On leaves of cultivated roses. December — June.
ERYSIPHE Hedw.
  E. COMMUNIS (Wallr.) Fr. On cultivated peas.
CAPNODIUM Mont.
  C. CÆSPITOSUM E. & E. Proc. Phil. Ac. Nat. Sc. 1894 p. 325. On loquat leaves.
  C. CITRI B. & D. On orange and lemon leaves.
EUROTIUM Link.
  E. HERBARIORUM (Wigg.) Lk. On various decaying substances.

Order Pyrenomyceteæ. Black Fungi.

ROSELLINIA Ces. & De Not.
  R. AQUILA (Fr.) De Not. On oak bark and grape stems.
CUCURBITARIA Gray.
  C. STENOCARPA E. & E. (n. sp. in lit.) On dead stems of Rhus diversiloba. June —
    September.
SPHÆRELLA Ces. & De Not.
  S. ARBUTICOLA Pk. On leaves of *Umbellularia californica.*
  S. SIDÆCOLA E. & E. Erythea 198. On leaves of *Sidalcea delphinifolia.* March —
    May.

STIGMATEA Fries.
S. GERANNII Fr. On leaves of *G. carolinianum*. February — April.

GNOMONIA Ces. & De Not.
G. ALNI Plowr. On leaves of *A. rhombifolia*. June — October.

OPHIOBOLUS Riess.
O. FULGIDUS (C. & P.) Sacc. On dead stems in Arroyo Seco. July.

CLYPEOSPHÆRIA Fckl.
C. HENDERSONIA (Ell.) Sacc. On dead stems.

PHYLLACHORA Nitch.
P. GRAMINIS (Pers.) Fckl. On *Muhlenbergia mexicana*. August — October.

NUMMULARIA Tul.
N. RUMPENS Cke. On oak bark and on sycamore wood. Jannary — April.

HYPOXYLON Bull.
H. CALIFORNICUM E. & E. (n. sp. in lit.) On dead stems of *Adenostoma fasciculatum*. June — September.
H. OCCIDENTALE Ell. & Ev. Proc. Phil. Ac. Nat. Sc. 1894, p. 345. On dead limbs and trunks. L. R. C. and S. G. Mts.

GLONIUM Muhl.
G. LINEARE (Fr.) Sacc. On old decorticated trunks of *Acer macrophyllum*.

HYSTEROGRAPHIUM Corda.
H. PROMINENS (P. & H) B. & G. On dead limbs.

ENDOCARPON (Hedw.) Fr.
E. MINIATUM Ach. On rocks.

Order Discomyceteæ. Lichens. Cup-fungi.

BUELLIA De Not., Tuck,
B. OIDALEA Tuck. On dead limbs.

CLADONIA Hoffm.
C. FIMBRIATA (L.) Fr. var. TUBÆFORMIS Fr. In soil in San Gabriel mountains.
C. FURCATA (Huds.) Fr. var. RACEMOSA Fl. Canyon sides.
C. PYXIDATA (L.) Fr. Canyon sides.

URCEOLARIA Tuck.
U. SCRUPOSA (L.) Nyl. On soil.

PERTUSARIA DC.
P. MULTIPUNCTA (Turn.) Nyl. On trees.

RINODINA Mass.
R. SOPHODES (Ach,) Nyl. On trees.

LECANORA Ach. Tuck.
L. PALLESCENS (L.) Schaer. On trees.
L. PALLIDA (Schaer. var. CANCRIFORMIS Tuck. On trees.
L. PRIVIGNA (Ach.) Nyl. On rocks.
L. SUBFUSCA (L.) Ach. On trees.

PLACODIUM (DC.) Naeg. & Hepp.
P. AURANTIACUM (Lightf.) Naeg. & Hepp. On rocks and trees.
P. BOLACINUM Truck. On rocks.
P. CERINUM (Hedw.) Naeg. & Hepp. var. PYRACEA Nyl. On trees.
P. FERRUGINEUM (Huds.) Hepp. On trees.
P. MURORUM (Hoffm.) DC. On rocks.

LEPTOGIUM Fr. Nyl.
L. ALBOCILIATUM Des. On trees and ground in A. S. and S. G. Mts.
L. PALMATUM (Huds.) Mont. Among moss in San Gabriel mountains.

COLLEMA Hoffm. Fr.
C. NIGRESCENS (Huds.). Ach. On trees.

PANNARIA Delis.
P. LANUGINOSA (Ach.) Koerb. On rocks and soil in San Gabriel Mts.

PELTIGERA (Willd. Hoffm.) Fee.
P. CANINA (L.) Höffm. On rocks and soil among moss.

STICTA (Schreb.) Fr.
S. PULMONARIA (L.) Ach. On rocks and trees.

40

UMBILICARIA Hoffm.
  U. PHAEA Tuck. On rocks.
PHYCIA (DC., Fr.) Th. Fr.
  P. HISPIDA (Schreb., Fr.) Tuck. On trees.
  P. OBSCURA (Ehrh.) Nyl. On trees.
  P. STELLARIS (Linn.) Tuck. On trees and rocks.
PARMELIA (Ach.) De Not.
  P. CAPERATA (Linn.) Ach. On trees and rocks.
  P. CONSPERSA (Ehrh.) Ach. On rocks,
  P. OLIVACEA (Linn.) Ach. On trees on Mount Lowe.
  P. PERLATA (Linn.) Ach.
  P. PHYSODES (Linn.) Ach. var. ENTEROMORPHA Tuck. On trees above 4000 feet.
  P. TILIACEA Hoffm. Floerk. On trees.
THELOSCHISTES Norm. Emend.
  T. CHRYSOPHTHALMUS (Lynn.) Norm. var. A. Com. On shrubs and trees.
  T. LYCHNEUS (Nyl.) Tuck. On trees.
USNEA (Dill.) Ach.
  U. BARBATA (Lynn.) Fr. var. ARTICULATA Ach. On trees in S. G. Mts.
EVERNIA Ach., Mann.
  E. VULPINA (Linn.) Ach. On coniferous trees above 3,500 feet.
CETRARIA (Ach.) Fr. Muell.
  C. CALIFORNICA Tuck. On trees.
RAMALINA Ach., De Not.
  R. CALCARIS (Linn.) Fr. var. FRAXINEA Fr. On trees.
  R. LAEVIGATA Fr. On trees.
  R. MENZIESII Tuck. On trees.
  R. RETICULATA (Noehd) Krem. On telephone poles.
RHYTISMA Fries.
  R. PUNCTATUM Pers. On leaves of *Acer macrophyllum.* July — October.
TAPHRINA Tul.
  T. DEFORMANS (Berk.) Tul. (Peach Curl.) On peach leaves. April — June.
CRYPTODISCUS Corda.
  C. ATROVIRENS (Fr.) Corda. On dried shrub in S. G. Mts. August.
LACHNELLA Fries. Cup-fungus.
  L. CONFUSA (Linn.) Fr. On decaying wood. January — April.
LACHNEA Fries. Cup-fungus.
  L. SCUTELLATA (Linn.) Fr. Com. On moist soil and decaying wood.
PSEUDOPEZIZA Fckel.
  P. MEDICAGINIS Sacc. On leaves of alfalfa.
  P. TRIFOLII (Bernh.) Fckl. On leaves of white clover.
PEZIZA Dill. Cup-fungus.
  P. CHRYSOCOMA (Bull.) C. & E. On decaying wood. February — April.
  P. SUBREPANDA C. & P. Moist soil at Oak Knoll. January — April.
  P. VIOLACEA (Gill.) Pers. Moist soil in Arroyo Seco. February — April.
  P. VESCICULOSA Bull. Frequent in moist soil. December — April.
HELVELLA Fries.
  H. CALIFORNICA Phil. Common under trees at Oak Knoll. January — April.
  H. CRISPA Fr. Under trees at Oak Knoll. April.
  H. LACUNOSA Afz. Frequent in shaded soil at Oak Knoll and A. S. Feb. — April.
MORCHELLA Dill. Morel.
  M. CONICA Pers. Occasional in shaded soil. February — May.

Order *Uredineæ. Rusts.*

UROMYCES Link.
  U. BETÆ Kuehn II. On cultivated beets.
  U. CARYOPHYLLINUS (Schrk.) Schrt. On garden pink (*Dianthus.*)
  U. CHORIZANTHIS Ell. & Hark. III. On *C. staticoides.* June — August.
  U. ERIOGONI (?) Ell. & Hark. II. On *E. elongatum* and *E. saxatile.* May — Oct.
  U. EUPHORBIÆ C. & P. II, III, On *E. serpyllifolia.* June — September.

U. JUNCI (Desm.) Tul. II, III. On *J. balticus, J. robustus, and J. xiphioides*. February — October.

U. LUPINI B. & C. I, II & III. On *L. albifrons, L. formosus bridgesii,* and *L. cytisoides*. March — July.

U. POLYGONI (Pers.) Fckl. II. On *P. aviculare*. June — August.

U. TEREBINTHI (DC.) Wint. II & III. On *Rhus diversiloba*. August.

U. TRIFOLII (A. & S.) Wint. I & III. On *T. gracilentum, T. macrei, T. microcephalum, T. ciliolatum* and *T. roscidum*. April — June.

U. ZYGADENUS (?) Pk. II. On *Z. fremonti*. March — April.

PUCCINIA Pers.

P. ACHYRODIS Diet. & Hol. (n. sp. in lit.) III. On *A. aureum*. June—August.

P. AMORPHÆ Curt. III. On *A. California* July — September.

P. BACCHARIDIS Diet. & Hol. Erythea 1: 250. II & III. On *B. viminea*. June—November.

P. CARICIS (Schum.) Rab. II & III. *C. barbaræ* and *C. filiformis latifolia*.

P. CLARKLÆ Pk. I & III. On *Zauschneria californica ;* III. On *Œnothera bistorta* and *Godetia bottæ*. May — July.

P. CONVOLVULI (Pers.) Cast. II & III. On *G. occidentalis*. June — August.

P. CORONATA Corda. II & III. On *Holcus lanatus*. May — September.

P. DIGITATA Ell. & Hark. III. On *Rhamnus crocea*.

P. EULOBII Diet. & Hol. Erythea 1:249. I & III. On *E. californicus*. April --Aug.

P. FLOSCULOSORUM (A. & S.) Roehl. III. On *Carduus californicus* and *C. occidentalis*. February — July.

P. GALII [Pers.] Schw. II & III. On *G. californicum, G. cinereum,* and *G. nuttallii*. May — September.

P. GILLÆ Hark. III. On *G. attractyloides*. June — August.

P. GRAMINELLA (Speg.) Diet. & Hol. III. On *Stipa eminens*. March — June.

P. GRAMINIS Pers. II & III. On *A. fatua, A. sativa. Elymus condensatus, E. triticoides,* and cultivated barley.

P. HARKNESSII Vize. III. On *Ptiloria cichoriacee*, June — August.

P. HELIANTHII Schw. II & III. On *H. annuus*.

P. HIERACII (Schum.) Mart. III. On *Malacothrix tenuifolia*. July.

P. INVESTITA Schw. I & III. On *Gnaphalium californicum*. July.

P. JONESII Pk. III. On *Vetæa arguta*. April —July.

P. MALVACEARUM Mont. III. On *M. parviflora* and hollyhock.

P. NODOSA Ell. & Hark. III. On *Brodiæa capitata*. April.

P. MCCLATCHIANA Diet. & Hol. Erythea II : 127. III. On *Scirpus sylvaticus microcarpus*.

P. MELLIFERA Diet. & Hol. Erythea 1:251. I & III. On *Salvia mellifera*. May — July.

P. MENTHÆ Pers. II & III. On *M. canadensis*. June — September

P. PALEFACIENS Diet. & Hol. Erythea II : 128, III. On *Arabis holbellii* February — April.

P. PIMPINELLÆ (Straus.)L. k. III. On *Osmorrhaza brachyopoda*. April — May.

P. POLYGONI-AMPHIBII Pers. II. On *P. acre*.

P. PROCERA Diet. & Hol. Erythea 1:249. II & III. On *Elymus condensatus, E. triticocoides,* and *E. americanus*.

P PRUNI-SPINOSÆ Pers. II & III. On leaves of cultivated plums, peaches and apricots. July — October.

P. PULVERULENTA Grev. III. *On Ephilobium paniculatum*. September.

P. RECONDITA Diet. &. Hol. Erythea 11: 128. III. On *Artemisia vulgaris californica*. September — April.

P. RUBIGO-VERA (DC.) Wint. II & III. On wheat *Bromus hookerianus* and *Hordeum murinum*. March — November.

P. TANACETI DC. II & III. On *Artemisia vulgaris californica*. July —November.

P. XANTHII Schw. III. On *X. strumarium*. August — October.

PHRAGMIDIUM Link.

P. SUBCORTICIUM (Schr.) Wint. On leaves and stems of *Rosa californica* and cultivated roses.

P RUBI-IDAEI (?) (Pers.) Wint. II. On leaves, stems, and fruit of cultivated blackberries. July — October.

ÆCIDIUM Pers. Cluster Cups.

Æ. CLEMATIDIS DC. On *C. ligusticifolia*. March — July.

Æ. EUPHORBIÆ Gmel. On *E. albomarginata* and *E. serpyllifolia*. July — September.

Æ. PHACELLÆ Pk. On *P. ramossissima*. February — April.
Æ. ROESTELIOIDES E. & E. On *Sidalcea delphinifolia*. March — May.
Æ. URTICÆ Schum. On *U. holosericea*. February — April.

UREDO Pers.
U. FILICUM (Lk.) Chev. On *Gymnogramme triangularis*. January — April.
U. PTERIDIS Diet. & Hol. Erythea II: 127. On *P. aquilina*. February.

MELAMPSORA Cast.
M. SALICIS-CAPRÆÆ (Pers.) Wint. II. On *S. lasiolepis*. June — August.

PUCCINIASTRUM Otth.
P. EPILOBII (Chaill.) Otth. II. On *E. adenocaulon occidentale*.

Order Ustilagineæ. Smuts.

USTILAGO Pers. Smit.
U. AUSTRO-AMERICANA Speg. On *Polygunum nodosum*. September — November.
U. AVENÆ (Pers.) Jens. On cultivated oats. April — June.
U. BROMINORA Fisch. de Wald. On *B. hookerianus*. April — June.
U. HORDEI (Pers.) K. & S. On cultivated barley. April — June.

IMPERFECT FUNGI.

Order Sphaeropsideæ.

PHYLLOSTICTA Pers.
P. FERAX Ell. & Ev. Proc. Phil. Ac. Nat. Sc. 1894. p. 355. On leaves of *Lupinus formosus bridgesii*. Sierra Madre. April.

DIPLODIA Fries.
D. UMBELLULARIÆ E. & E. (n. sp. in lit.) On *U. californica*. July —October.

HENDERSONIA Mont.
H. UMBELLULARIÆ E. & E. (n. sp. in lit.) On *U. californica*. July — October.

ASCOCHYTA Lib.
A. GRAMINICOLA Lib. On *Poa Annua*. December—March.
A. MENZIESII Ell. & Ev. (n. sp. in lit.) On leaves of *Arbutus menziesii*. S. G. Mts. February.

ACTINONEMA Fries.
A. ROSÆ (Lib.) Fr. On leaves of *R. californica*.

SEPTORIA Fries.
S. ALNIFOLIA E. & E. On leaves of *A. rhombifolia*. April—November.
S. DULCAMARÆ Desm. On leaves of *Solanum douglasii*. January—May.
S. FUMOSA Pk. On leaves of *Solidago californica*. May—October.
S. IRREGULARIS Pers. On leaves of *Rhus diversiloba*. July—October.
S. MIMULI Ell & Kell. On leaves of *M. glutinosus*. June—August.
S. ŒNOTHERÆ Westd. On leaves of *Œ. biennis*. June—October.
S. POLYGONORUM Desm. On leaves of *P. nodosum*. June—November.
S. RHAMNI-CATHARTICÆ Ces. On leaves of *R. californica*.
S. RUBI Westd. On leaves of *R. ursipus*.
S. SCROPHULARIÆ Pk. On leaves of *S. californica*. January—May.

SPHAEROPSIS Lev.
S. ALNI C. & E. On bark of *A. rhombifolia*.

ENTOMOSPORIUM Lev.
E. MACULATUM Lev. On leaves of *Heteromeles arbutifolia*.

AMEROSPORIUM Speg.
A. CINCTUM Ell. & Ev. (n. sp. in lit.) On stems of Gladiolus. July.

Order Melonconieæ.

GLOEOSPORIUM Desm. & Mont.
G. CERCOCARPI Ell. & Ev. Erythea 11:25. On leaves of *C. parvifolius*. June—August.
G. PHYLLACHOROIDES Ell. & Ev. Erythea 1:201. On leaves of *Artemisea vulgaris californica*.

CYLINDROSPORIUM Unger.
    C. CEANOTHI Ell & Ev. On leaves of *C. divaricatus*. June — August.
    C. TOXICODENDRI (Curt.) Ell & Ev. Proc. Phil. Ac. Nat Sc. VII: 160. On leaves of *Rhus diversiloba*. July — October.
MELANCONIUM Link.
    M. ACERINUM Ell. & Ev. Proc. Phil. Ac. Nat. Sc. 1891, p. 373. On dead limbs of *A. macrophyllum*. February and April.
    M. BICOLOR Nees. On bark of *Alnus rhombifolia*. July — October.
MARSONIA Fisch.
    M. POTENTILLAE (Desm.) Fisch. On leaves of *P. californica*. March — July.

Order *Hyphomyceteae.*

OIDIUM Link.
    O. ERYSIPHOIDES Fr. On *Phacelia ramossissima*, *P. whitlavia*, *Stachys californica*, *Heterotheca grandiflora*, *Anthemis cotula*, *Galium aparine*, *Artemisia vulgaris californica*, and *Helianthus annuus*.
    O. MONILIOIDES Lk. On *Bromis americanus*.
TRICHODERMA Pers.
    T. LIGNORUM (Tode.) Hartz. On decaying wood. January — April.
BOTRYTIS Michx.
    B. VULGARIS Fr. On withered rose petals.
PENICILLIUM Lk.
    P. GLAUCUM Lk. On all kinds of decaying matter.
RAMULARIA Ung.
    R. DECIPIENS Ell & Ev. On leaves of *Rumex crispus*.
    R. MELILOTI Ell. & Ev. Erythea II : 26. On *M. indica*. Oct.
    R. URTICAE Ces. On *U. holosericea*.
CERSPORELLA Sacc.
    C. PROLIFICANS E. & E. On *Sambucus glauca*. April — November.
TORULA Pers.
    T. SPORIDESMOIDES Ell. & Ev. Proc. Phil. Ac. Nat. Sc. 1894, p. 377. On bark of dead limbs.
HORMISCIUM Kunze.
    H. STILBOSPORA (Corda) Sacc. On decaying wood.
BACTRIDIUM Kunze.
    B. ELLISII Beck. On decaying wood. February — April.
DEMATIUM Pers.
    D. VINOSUM Mass. On culture media in laboratory.
SCOLECOTRICHUM Kunze G Schm.
    S. ASCLEPIADIS Ell. & Ev. Erythea I : 203. On leaves of *A. eriocarpa*. June — November.
    S. GRAMINIS Fckl. On cultivated barley and oats.
FUSICLADIUM Bon.
    F. DEPRESSUM B. & Br. On *Velaea arguta*. June — August.
CLADOSPORIUM Link.
    C. AROMATICUM E. & E. (n. sp. in lit.) On leaves of *Rhus aromatica*. July — Aug.
    C. CARPOPHILUM Thm. On ripe apricots. July — August.
    C. EPIPHYLLUM Pers. On leaves of *Eucalyptus globulus*.
    C. HERBARUM (Pers.(Link. On various parts of decaying plants.
    C. PAEONAE Sacc. On leaves of *P. californica*. May—August.
    C. TYPHARUM Desm. On *T. latifolia*.
CERCOSPORA Fries.
    C. BETICOLA Sacc On beet leaves.
    C. CIRCUMCISSA Sacc. On leaves of *Prunus illicifolia*.
    C. CLAVICARPA E. & E. Erythea 11:26. On *Pilosia virgata*. July—September.
    C. CROTONIS E. & E. On leaves of *C. californicus*. June—September.
    C. EPILOBII Schn. On leaves of *E. adenocaulon occidentale*. July—September.
    C. HETEROMELES Hk. On leaves of *H. arbutifolia*. June—November.
    C. NASTURTII Pass. On leaves of *N. officinale*. July—September.

41

C. ROSICOLA Pass.   On leaves of cultivated roses and *R. californica.* July—
September.
C. SAURURI E. & E.   On leaves of *Anemopsis californica.* June—October.
C. SQUALIDA Pk.   On leaves of *Clematis ligusticifolia.* July—November.
C. VIOLE Sacc.   On leaves of cultivated violets.
C. VITICOLA (Ces., Sacc.   On leaves of *V. californica* and cultivated grapes. June
—October.
HETEROSPORIUM Klotzsch.
H. EUCALYPTI E. & E.   Proc. Phil. Ac. Nat. Sc. 1894, p. 381.   On *E. globulus*
leaves. November.
H. EUCALYPTI E. & E.   var. MACULICOLUM (n. var. in lit.)   On leaves of *Eriodic-
tron californicum.* June—August.
H. PHRAGMITES (Opiz) Klot. On corn leaves. September—November.
CONOTHECIUM Corda.
C. UMBELLULARLE E. & E. (n. sp. in lit.) On stems of *U. californica.* June—
October.
MACROSPORIUM Fries.
M. CAUDATUM C. & E.   On dead leaves of *Vitis californica.* August.
M. MAYDIS C. & E.   On corn leaves.
M. PELARGONII Ell & Ev. Proc. Phil. Ac. Nat. Sc. 1894, p. 383. On leaves of
cultivated Geranium. February.
M. SOLANI Ell. & Mart. On potato leaves.
STEMPHYLIUM Wallr.
S. MACROSPOROIDEUM B. & Br. On culture media in laboratory.
S. ALTERNARLE Cke. On culture media in laboratory.
EPICOCCUM Lk.
E. NEGLECTUM Desm. On corn leaves. September—November.
PODOSPORIELLA Ell. & Ev. Proc. Phil. Ac. Nat. Sc. 1894, p. 385.
P. HUMILIS E. & E. Ibid. On leaves of *Garrya veatchii.* June—October.

CLASS III. BASIDIOMYCETES. Puff-Balls, Toadstools, Pore-Fungi.

Order Gasteromycetea.   Puff-balls.

PHALLUS Mich. Stink-horn.
P. RAVENELII B. & C. Frequent in rich soil and in lawns. October—March.
SECOTIUM Mont.
S. DECIPIENS Pk. (n. sp. in lit.) Common along gutters.
GEASTER Mich. Earth-star.
G. LIMBATUS Fr. O. K.; A. S. under trees. February—April.
G. MINIMUS Schw. Oak Knoll. February—April.
G. SACCATUS Fr. Arroyo Seco. March.
ASTRÆUS Morg. Hygrometric Earth-star.
A. HYGROMETRICUS (Pers.) Morg. Com. under trees. January—March.
TYLOSTOMA Pers. Stalked puff-ball.
T. CAMPESTRE Morg. Along Wilson Trail and in dry soil at Devil's Gate. Janu-
ary—July.
CALVATIA Fr. Puff-ball.
C. CÆLATA Bull. In rich soil. February—April.
C. FRAGILIS Vitt. In lawn and along Wilson Trail, 4,000 feet. January—October.
C. HESPERIA Morg. (n sp. in lit.) Along streets and in yards. November—De-
cember.
LYCOPERDON Tourn. Puff-ball.
L. CELESFORME Bull. Oak knoll. January—March.
L. MOLLE Pers. Along street. December—March.
CATASTOMA Morg.
C. CIRCUMCISSUM (B. & C.) Morg. Along street. February—March.
BOVISTA Dill. Puff-ball.
B. AMMOPHILA Lev. Com. below 2,000 feet. December—March.
SCLERODERMA Pers.
S VULGARE Fr Frequent in rich soil. September—January.

CYATHUS Hall.
    C. VERNICOSUS (Bull.) DC. On decaying wood and in soil. March—May.
CRUCIBULUM Tul. Bird-nest fungus.
    C. VULGARE Tul. On decaying wood and in soil in and near the A. S. February
    —April.

Order—Hymenomycetes, Toadstools, Pore-Fungi, etc.

AMANITA Fries.
    A. PHALLOIDES Fr. Com. under oak trees. February—April. Poisonous.
AMANITOPSIS Roze.
    A. VELOSA Pk. (n. sp. in lit.) Com. under oaks. February—May. Edible.
LEPIOTA Fries.
    L. ANGUSTANA Britz. Frequent under trees. December—February.
    L. FULVODISCA Peck. Tor. Bull. 22. 198. Among moist leaves in Arroyo Seco.
    January.
    L. NAUCINOIDES Pk. Abundant in lawns. August—November. Edible.
ARMILLARIA Fries.
    A. MELLEA Vahl. Com. at base of trees and stumps. September—January. Edi-
    ble.
    A. MELLEA Vahl. Var: NIGRIPES Pk. (n. var. in lit.) On willow trunk in W. C.
    December.
TRICHOLOMA Fries. Pers.
    T. EQUESTRE Linn. Among leaves under oaks. February.
    T. MELALEUCUM Peri. Under oaks in W. C. January.
    T. NUDUM Bull. Frequent under trees. January—March.
    T. RUSSULA Schaef. Under oaks in O. K. C. February.
COLLYBIA Fries.
    C. ALBOGRISEA Pk. Tor. Bull. 22:199. Frequent in W. C. January and February.
    C. DRYOPHILA Bull. Frequent among leaves under trees. January and February.

A. J. McClatchie, del.

MYCENA ACICULA.
Two-thirds natural size.

MYCENA Fries.
    M. ACICULA Schaef. On decaying wood. February.
    M. ATROALBOIDES Pk. Abundant among wet leaves. December—February.
    M. ELEGANTULA Pk. Tor. Bull. 22:199. Abundant among wet leaves. December—February.

RUSSULA Fries.
    R. EMETICA Fr. Com. under oaks. November—March. Poisonous.

LACTARIUS Fries.
    L. CAMPHORATUS Fr. Com. under oaks. January—March.
    L. INSULSUS Fr. Com. under oaks. January—March.

CLITOCYBE Fries.
    C. PUSILLA Pk. Tor. Bull. 22:199. On manure. February.
    C. TORTILIS Bolt. In moist soil in A. S. February.

OMPHALIA Fries.
    O. PYXIDATA Bull. On moist banks. December—February.
    O. PYXIDATA Bull. var. FURCATA Pk. (n. var. in lit.) Com. in waste soil. January—March.

PLEUROTUS Fries.
    P. OSTREATUS Jacq. (Oyster mushroom.) On decaying wood. January—March. Edible.
    P. SAPIDUS Kalch. On decaying wood. January—March. Edible.

HYGROPHORUS Fries.
    H. ERUBESCENS Fr. Com. under oaks. December—January.

LENZITES Fries.
    L. BETULINA Fr. On oak stumps in M. C. and in W. C. January—March.

SCHIZOPHYLLUM Fries.
    S. COMMUNE Fr. Com. on decaying logs.

VOLVARIA Fries.
    V. SPECIOSA Fr. Com. in grain fields and waste ground. Jan.—March. Edible.

ENTOLOMA Fries.
    E. FERUGINANS Pk. Tor. Bull 22:200. Abundant under oaks. January—April. Edibles.

LEPTONIA Fries.
    L. EDULIS Pk. Tor. Bull. 22:201. Frequent in grass and among weeds. December—February. Edible.

ECCILIA Fries.
    E. NIGRICANS Pk. Tor. Bull. 22:202. Abundant in waste soil. December—March. Edible.

PHOLIOTA Fries.
    P. ANOMALA Pk. Tor. Bull. 22:202. Among leaves in A. S. January—March.
    P. PRAECOX Pers. In grass. March—June.

PLUTEOLUS Fries.
    P. LUTEUS Pk. Tor. Bull. 22:203. Among leaves and grass under trees. February—March.

BOLBITIUS Fries.
    B. FRAGILIS Fr. var. ALBIPES Pk. (n. var. in lit.) In grass. January.
    B. TENER Berk. In lawns. April—June.

HEBELOMA Fries.
    H. FOEDATUM Pk. Tor. Bull. 22:202. In grass along streets. January—March.
    H. ISCHNOSTYLUM Cke. Among leaves in the A. S. January—March.
    H. MESOPHORUM Fr. Among moss in A. S. December—March.

NAUCORIA Fries.
    N. MELINOIDES Fr. In grass. March—May.
    N. SEMIORBICULARIS Bull. Com. among grass and weeds. February—June.

GALERA Fries.
    G. LATERITIA Fr. In lawns. April—June.
    G. OVALIS Fr. In rich soil. February—June.
    G. TENERA Schaeff. In lawns. Very common.

TUBARIA Smith.
    T. PALLESCENS Pk. Tor. Bull. 22:202. On leaves and sticks under trees. January—March.

A. J. McClatchie del.

PLUTEOLUS LUTEUS.
One-half natural size.

CREPIDOTUS Fries.
  C. HEPATIZON Berk. On decaying stumps in A. S. December—March.
  C. HERBARUM Pk. On leaves and sticks in A. S. January—March.
  C. SUBVERANTUS Pk. (n. sp. in lit.) Com. on leaves and sticks. December—April.

CORTINARIUS Fries.
  C. VIRGATUS Pk. Tor. Bull. 22:203. Among leaves under trees at O. K. February.

AGARICUS Linn.
  A. CALIFORNICUS Pk. Tor. Bull. 22:203. Com. under oaks. December—February. Edible.
  A. CAMPESTER L. (Common mushroom.) In forging shop and in gardens. Edible.

STROPHARIA Fries.
  S. BILAMELLATA Pk. Tor. Bull. 22:204. Along street. January. Edible.
  S. SEMIGLOBATA Batsch. In lawns. April—June.
  S. STERCORARIA Fr. On manure in woods. December—March.

HYPHOLOMA Fries.
  H. APPENDICULATUM Bull. Among leaves in W. C. January—February.
  H. FASCICULARE Hud. At base of stumps, A. S., R. C. January—March.
  H. INCERTUM Pk. In soil under trees. December—February.
  H. LONGIPES Pk. Tor. Bull. 22:204. Among wet leaves in cañons. January—March.
  H. MADRODISCUM Pk. Among leaves in in L. R. C. January—February.
  H. PERPLEXUM Pk. On rotten wood in R. C. March.

PSATHYRELLA Fries.
  P. DISSEMINATA Pers. At the base of stumps. December—March.

PANAEOLUS Fries.
  P. DIGRESSUS Pk. Tor. Bull. 22:205. On manure. July.

P. INTERMEDIUS Pk.   Tor. Bull. 22:205.   In sandy soil along streets and in washes.
   January—March.
P. RETIRUGIS Fr.   In rich soil.   April—June.
P. SUB-BALTEATUS B. & Br.   Common in lawns.   April—June.

COPRINUS Pers.   (Dissolving Toadstools.)
   C. CALYPTRATUS Pk.   In rich cultivated soil.   March—May.
   C. COMATUS Fr.   (Shaggy-maned mushroom.)   In rich soil.   Edible.
   C. CONGREGATUS Bull.   In rich soil.   March.
   C. LAGOPUS Fr.   In manure.   January—March.
   C. MICACEUS Fr.   At base of trees and stumps.   December—April.
   C. PLICATILIS Fr.   In rich soil.   March.
   C. PLUMBEUS Pk.   In lawn.   June—August.
   C. RADIATUS Fr.   On manure.   March.

BOLETUS Dill.   Stalked Pore-fungus.
   B. SUBTOMENTOSUS L.   Com. under trees.   December—April.

POLYPORUS Mich.   Pore-fungus.
   P. ADUSTUS (Willd.) Fr.   On decaying logs and stumps.   December—March.
   P. BULBIPES Fr.   San Gabriel Mts.   March—May.
   P. DENDRITICUS Fr.   On decaying boards.   January—March.
   P. GILVUS Schw.   On decaying oak.
   P. LEUCOMELAS Fr.   San Gabriel Mts.   March—May.
   P. LEUSCULUM B. & C.   On decaying oak.
   P. ROSEUS A. & S.   On decaying *Pseudotsuga* in San Gabriel Mts.
   P. SCRUPIRUS Fr.   On decaying wood.
   P. SULPHUREUS (Bull.) Fr.   On decayed wood.   February—April.

FOMES Fries.
   F. APPLANATUS (Pers.) Wallr.   On decaying oaks.   December—May.
   F. IGNIARIUS (L.) Fr.   On decaying oak.
   F. LUCIDUS (Leys.) Fr.   On decaying oak.   January—April.

POLYSTICTUS Fries.
   P. HIRSUTUS Fr.   On decaying wood.   January—June.
   P. PERGAMENUS Fr.   On decaying *Pseudotsuga* in San Gabriel Mts.
   P. VERSICOLOR (L.) Fr.   On decaying wood.   December—April.

HYPHOLOMA FASCICULARE.
Two-fifths natural size.

TRAMETES Fries.
    T. PECKII Kalch. On decaying wood.
FAVOLUS Fries.
    E. PURPURASCENS B. & C. On decaying wood in San Gabriel Mts.
MERULIUS Hall.
    M. CORIUM Fr. On decaying oak. December—May.
    M. LACHRYMANS Fr. On moist soil and decaying wood. January—April.
PHLEBIA Fries.
    P. MERISMOIDES Fr. On decaying bark. January—April.
RADULUM Fries.
    R. ORBICULARE Fr. On decaying bark. January—April.
HYDNUM Linn.
    H. OCHRACEUM Pers. On decaying oak bark. March—May.
STEREUM Pers.
    S. ALBOBADIUM Schw. On burnt and decaying wood and bark.
    S. GALEATII Berk. On oak bark.
    S. HIRSUTUM (Willd.) Fr. On oak bark. Common.
    S. MOLLE Lev. On oak bark.
    S. PURPUREUM Pers. On decaying wood and bark.
    S. SPADICEUM Fr. On oak bark.
    S. TRISTE B. & C. On burnt wood.
    S. VERSICOLOR (Schw.) Fr. Com. on oak bark.
CORTICIUM Fries.
    C. CARNEUM B. & C. On decaying wood.
    C. CINEREUM (Pers.) Fr. On stems of *Adenostoma.*
    C. COMEDEUS (Nees.) Fr. On decaying oak branches.
    C. LACTEUM Fr. On decaying sycamore. January—April.
    C. SEBACEUM (Cke.) Mass. On moss. January.
HYMENOCHAETE Lev.
    H. RUBIGINOSA (Schr.) Lev. On decaying roots and stumps.
PENIOPHORA Cooke.
    P. MORICOLA Berk. On dry wood.
    P. OCHRACEA (Fr.) Mass. On decaying sycamore.
    P. QUERCINA (Fr.) Cke. On oak bark.
CONIOPHORA D. C.
    C. FUMOSA Pers. On sycamore bark.
CLAVARIA Vaill.
    C. CORALLOIDES Linn. Among leaves at O. K. January—March.
    C. GRISEA Pers. Among leaves at O. K. January—March.
    C. INAEQUALIS Berk. Among leaves at O. K. January—March.
    C. LIGULA Fr. Under oaks at O. K. January—March.
TREMELLA Dill.
    T. AURANTIA Schw. On decaying wood. December—March.
    T. INFLATA Fr. On rotton wood. March.
EXIDIA Fries.
    E. GLANDULOSA Fr. On dead oak branches. December—March.

CLASS IV. RHODOPHYCEÆ.

Order Floridea.

BATRACHOPERMUM Roth.
    B. GELATINOSUM (L.) Woods. Stream in Arroyo Seco.
    B. VAGUM Ag. Streams in A. S. and M. C.

BRYOPHYTA.

CLASS I. HEPATICÆ. Liverworts.

Order Ricciaceæ.

RICCIA Linn.
  R. AGGREGATA Und. Bot. Gaz. 19:275. Com. on bank and trodden soil. December—April.
  R. ARVENSIS Aust. var. HIRTA Aust. Com. on banks and in gravelly soil below 3000 feet. December—April.
  R. GLAUCA Linn. Shaded soil. January—April.
SPILEROCARPUS Mich.
  S. TERRESTRIS Mich. var. CALIFORNICUS Aust. Com. on moist compact soil. January—April.

Order Marchantiaceæ.

MARCHANTIA Linn.
  M. POLYMORPHA Linn. Wet soil. O. K. ; A. S. ; Wilson Trail.
GRIMALDIA Raddi.
  G. CALIFORNICA Gottsche. Shaded soil, Wilson Trail, 4000 feet. March—June.
CRYPTOMITRIUM Aust.
  C. TENERUM Aust. Moist, shaded soil in canyons, below 3000 feet. March—June.
ASTERELLA Pal. de Beauv.
  A. CALIFORNICA (Hampe.) Und. Com. on shaded soil below 5000 feet. November—April.
  A. NUDATA (Howe) Und. Frequent in gravelly soil below 2,000 feet. February—April.
AYTONIA Forst.
  A. ERYTHROSPERMA (Sulliv.) Und. Wet soil. R. C. Wilson Trail, 5,000 feet. March—May.
LUNULARIA Mich.
  L. CRUCIATIA (Linn.) Dum. Com. in green-houses.
TARGIONIA Mich.
  T. HYPOPHYLLA Linn. Com. on shaded banks below 4,000 feet. December—April.

Order Anthocerotaceæ.

ANTHOCEROS Linn. Horned-liverwort.
  A. FUSIFORMIS Aust. Moist soil. Not common. March—May.
  A. LAEVIS L. Frequent on wet rocks along streams. January—September.

Order Jungermaniaceæ Scale-mosses.

FOSSOMBRONIA Raddi.
  F. LONGISETA Aust. Com. on shaded banks. December—April.
FRULLANIA Raddi. Scale-moss.
  M. BOLANDERI Aust. Bark on trees in S. G. Mts. December—June.
MADOTHECA Dumort. Scale-moss.
  M. BOLANDERI Aust. Com. on rock in S. G. Mts. December—April.

CLASS II. MUSCI.

Order Bryaceæ.   True Mosses.

EUCLADIUM Br. & Sch.
  E. VERTICILLATUM (Linn.) B. & S. Shaded soil in S. G. Mts. None bearing sporogonia collected.
DICRANOWEISIA Lindb.
  D. CIRRHATA (Hedw.) Lindb. Shaded soil in S. G. Mts. March—May.
FISSIDENS Hedw.
  F. GRANDIFRONS Brid. Under swiftly running water in canyons S. G. Mts.
  F. LIMBATUS Sulliv. Com. in shaded soil. February—May.

BARBULA Hedw.
B. MUELLERI B. & S. Frequent on rocks. February—May.
B RURALIS (Linn.) Hedw. Com. on rocks and soil. February—May.
B. VINEALIS Brid. Frequent in S. G. Mts. and A. S. March—June.

GRIMMIA Ehrh.
G. CALIFORNICA Sulliv. On dry rocks in S. G. Mts. March—June.
G. LEUCOPHÆA Grev. Com. on rocks and bark of trees. February—May.
G. TRICHOPHYLLA Grev. Com. on rocks and bark of trees. February—May

HEDWIGIA Ehrh.
H. CILIATA (Dicks.) Ehrh. Frequently on rocks in S. G. Mts. April—July.

ORTHOTRICHUM Hedw.
O. LYELLII H. & T. Com. on bark of trees, especially in S. G. Mts. March—May.

FUNARIA Schreb.
F. HYGROMETRICA (Linn). Sibth. Com. below 3,000 feet. December—March.

BARTRAMIA Hedw.
B. MENZIESII Turn. Com. in S. G. Mts. February—April.

LEPTOBRYUM Schimp.
L. PYRIFORME (Linn.) Schimp. Com. in shaded soil below 3,000 feet. February—April.

WEBERA Hedw.
W. NUTANS (Schreb.) Hedw. Shady canyon sides and decaying logs. Feb.—April.

BRYUM Dill.
B. ARGENTEUM Dinn. Com. below 2,000 feet. February—May.
B TORQUESCENS B. & S. Frequent below 4,000 feet. March—June.
B. TURBINATUM Schw. Foothills S. G. Mts. March—May.

AULACOMNIUM Schwaegr.
A. ANDROGYNUM (Linn.) Schw. Near Strain's Camp, Wilson's Peak. April—May.

POLYTRICHUM Linn.
P. PILIFERUM Schreb. Echo Mt. No specimens bearing sporogonia found.

ALSIA Sulliv. Feather moss.
A. ABIETINA Sull. Com. on rocks and trees in S. G. Mts. April—May.
A. LONGIPES S. & L. Frequent on rocks in S. G. Mts. None bearing sporogonia collected.

PTEROGONIUM Swartz.
P. GRACILE (Linn.) Swartz. Rocks in canyons S. G. Mts. None bearing sporogonia collected.

ANTITRICHIA Brid.
A. CALIFORNICA Sull. Wet rocks in Rubio Canyon. None bearing sporogonia collected.

CLAOPODIUM Dill.
C. LEUCONEURUM S. & L. Rubio Canyon. None bearing sporogonia collected.

CAMPTOTHECIUM Dill.
G. ARENARIUM (Lesq.) Ren. & Card. On dry rocks and soil. February—May.
C. PINNATIFIDUM S. & L. Com. on rocks in A. S. and canyons of S. G. Mts.

BRACHYTHECIUM Schrimp.
B. RUTABULUM S. & L. Canyons of S. G. Mts. March—May.

SCLEROPODIUM Schrimp.
S. CAESPITOSUM (Wils.) Br. & Sch. On dry rocks in A. S. and R. C. March—June.
S. ILLECEBRUM (L.) Br. & Sch. Canyons of S. G. Mts. March—May. Common.
S. OBTUSIFOLIUM (Hook) Kindb. On rocks under running water in R. C. None bearing sporogonia collected.

EURYNCHIUM Schrimp.
E. STOKESII (Turn.) B. & S. Rubio Canyon. None bearing sporogonia collected.

RHYNCHOSTEGIUM Schrimp.
R. SERRULATUM Hedw. Rubio Canyon. None bearing sporogonia collected.

42

AMBLYSTEGIUM Schrimp.
    A. IRRIGUM Hook. & Wils. Moist soil in R. C. None bearing sporogonia collected.
    A. SERPENS (Linn.) B. & S. Com. along streams. February—May.

PTERIDOPHYTA.

CLASS I. FILICINÆ.

Order Filices.   Ferns.

CYSTOPTERIS Bernh. Bladder-fern.
    C. FRAGILIS Bernh. L. St. A. C., Rare. January—July.
ASPIDIUM Swz. Shield-fern.
    A. MUNITUM Kaulf. M. C., R. C., Eaton C., Mt. L., W. Pk.
    A. RIGIDUM Swz. var. ARGUTUM Eat. A. S., S. G. Mts., Com.
ASPLENIUM Linn. Spleenwort.
    A. TRICHOMANES Linn. var. INCISUM Moore. M. C., L. St. C. Not. common.
WOODWARDIA Smith. Chain-fern.
    W. RADICANS Sm. O. K., A. S., M. C., R. C., Eaton C., L. St. A. C.
PELLÆA Link. Cliff brake.
    P. ANDROMEDÆFOLIA Fee. (Coffee-fern.) O. K., A. S., S. G. Mts. Com.
    P. ORNITHOPUS Hook. (Bird-foot-fern.) A. S., S. G. Mts. Com.
PTERIS Linn. Brake.
    P. AQUILINA Linn. (Common brake.) O. K., A. S., M. C., R. C. Com. January—October.
ADIANTUM Linn. Maidenhair.
    A. CAPILLUS-VENERIS. Linn. (Venus' Hair.) M. C., R. C., E. C. Abundant.
    A. EMARGINATUM Hook. O. K., A. S. Abundant. January—June.
    A. PEDATUM Linn. L. St. A. C. Rare.
CHEILANTHES Swz.
    C. CALIFORNICA Mett. (Lace-fern.) A. S., S. G. Mts. Com.
    C. MYRIOPHYLLA Desv. Mt. L., W. Pk. Com. above 4,000 feet. December—July.
NOTHOLÆNA R. Br.
    N. NEWBERRYI Eat. (Cotton-fern.) Echo Mt. Rare.
GYMNOGRAMME Desf.
    G. TRIANGULARIS Kaulf. (Golden-back fern.) O. K., A. S., S. G. Mts. December—April.
POLYPODIUM Linn. Polypody.
    P. CALIFORNICUM Kaulf. A. S., S. G. Mts. Com December—June.

Order Salviniaceæ.

AZOLLA Lam.
    A. FILICULOIDES Lam. O. K., A. S. Abundant. June—August.

CLASS II. EQUISITINÆ.

Order Equisetaceæ.   Horse-tails.

EQUISETUM Linn. Horse-tail. Scouring-rush.
    E. MEXICANUM Milde. A. S., M. C: January—March.
    E. ROBUSTUM A. Br. A. S., M. C., R. C., E. C.
    E. TELMATEIA Ehrh. O. K., A. S. Abundant. January—October.

CLASS III. LYCOPODINÆ. CLUB-MOSSES

Order Selaginellaceæ.

SELAGINELLA Beauv. Little Club-moss.
    S. RUPESTRIS (L.) Spring. A. S., S. G. Mts. Common. January—March.

SPERMAPHYTA.

CLASS I. GYMNOSPERMÆ.

Coniferæ. Conifers.

PINUS Tourn. Pine.
  P. ALBICAULIS Eng. W. Pk. April—June.
  P. MONOPHYLLA T. & F. (Nut Pine.) Mt. Lowe. A single specimen.*
  P. MONTICOLA Dougl. W. Pk. April—June.
  P. PONDEROSA Dugl. W. Pk. April June.
PSEUDOTSUGA Carr. False Hemlock.
  P. MACROCARPA (Torr.) Lem. S. G. Mts. Com. April—June.
LIBOCEDRUS Endl. White Cedar.
  L. DECURRENS Torr. W. Pk. April—June.
JUNIPER Linn. Juniper.
  J. CALIFORNICA Carr. Near Sierra Madre. March—May.

CLASS II. ANGIOSPERMÆ.

Sub-Class I. Monocotyledones.

Typhaceæ.

TYPHA Tourn. Cat Tail.
  T. ANGUSTIFOLIA Linn. O. K. March—June.
  T. LATIFOLIA Linn. O. K., A. S. March—June.

Naiadaceæ.

POTAMOGETON Tourn. Pond-weed.
  P. PECTINATUS Linn. O. K. April—June.
  P. PUSILLUS Linn. O. K. April—June.
ZANNICHELLIA Presl.
  Z. PALUSTRIS Linn. O. K. April—June.

Gramineæ. Grasses.

ANDROPOGON Linn.
  A. MACROURUS Michx. R. C., E. C. August—October.
  A. SORGHUM Brot. (Sorghum.) Occasionally escaped, and often persisting in old
    sorghum fields.
PASPULUM Linn.
  P. DISTICHUM Linn. (Knot-grass.) Com. shaded soil. August—October.
PANICUM Linn.
  P. CAPILLARE Linn. (Old witch grass.) Along streets and in waste soil. July—
    October.
  P. CRUS-GALLI Linn. (Barn grass.) Along streets and in waste soil. July—Sept.t
  P. SANGUINALE Linn. (Crab grass.) Along streets and in waste soil. July—Sep.
SETARIA Beauv.
  S. GLAUCA (Linn.) Beauv. (Fox-tail grass.) In cultivated soil. June—September.
PHALARIS Linn.
  P. INTERMEDIA Bosc. (Canary grass.) Waste places in Pasadena. March—June.

---

*This is the only tree of the Indian nut pine known to exist on the front or middle ranges of our Pasadena mountains, and it has a history. On October 10, 1887, Jason and Owen Brown built a cairn on this mountain top. (See page 369, "Mt. Lowe.") They noticed this rare tree, with its roots so much denuded by rain-wash and wind that it was ready to die; and they gathered and brought soil in their little tin dinner pail to pack around its exposed roots, thus saving its life at that time, and hence it has been called the "Osawatamie pine tree." Then on August 15, 1893, Dr. Reid and wife found it perishing again from the same causes; and Mrs Reid gathered loose dirt and mulch from between rocks and dragged it on an old barley sack which she had found, to the roots of the tree, while the Doctor laid up a wall of rocks on the lower side to hold the dirt in place; and so its life was saved again. They also broke off some of its dead branches, to give the live part a better chance. The tree was then ten or twelve feet high. There are said to be some trees of the same species on San Gabriel peak, but this is not yet verified by competent testimony.—ED.

ARISTIDA Linn.
    A. DIVARICATA HBK. Altadena. June—September.
STIPA Linn. Feather grass.
    S. CORONATA Thurb. S. G. Mts. June—September.
    S. EMINENS Cav. O. K., S. G. Mts. April—July.
    S. SETIGERA Presl. O. K., S. G. Mts. April—June.
MUHLENBERGIA Trin.
    M. DEBILIS Trin. Arroyo Seco. May—July.
    M. MEXICANA (Linn.) Trin. A. S., M. C., R. C., E. C. June—September.
PHLEUM Linn.
    P. PRATENSE Linn. (Timothy.) Along streets and in lawns. June—September.
SPOROBOLUS R. Br.
    S. AIROIDES (Steud.) Torr. Along streets. July—September.
EPICAMPES Presl.
    E. RIGENS Benth. Arroyo Seco. June—August.
POLYPOGON Desf.
    P. LITTORALIS Smith. Moist soil in Lincoln Park. May—July.
    P. MONSPELIENSIS Desf. Moist soil about Pasadena. April—July.
AGROSTIS Linn. Bent grass.
    A. ATTENUATA Vasey. R. C. June—August.
    A. EXARATA Trin. O. K., A. S. May—June.
    A. MICROPHYLLA Steud. Altadena, Sierra Madre. June—August.
    A. VERTICILLATA Vill. O. K., R. C. June—August.
GASTRIDIUM Beauv.
    G. AUSTRALE Beauv. Altadena, Sierra Madre. June—August.
HOLCUS Linn.
    H. LANATUS Linn. O. K., Santa Anita. June—August.
TRISETUM Beauv.
    T. BARBATUM Steud. Arroyo Seco. May—June.
AVENA Linn. Oats.
    A. BARBATA Steud. (Wild oats.) Along streets. March—July.
    A. FATUA Linn. (Wild oats.) Com. below 2,000 feet. February—June.
CYNODON Rich.
    C. DACTYLON Pers. (Bermuda grass.) Along streets and in lawns. March—July.
ERAGROSTIS Beauv.
    E. MEXICANA Trin. Along streets. August—October.
MELICA Linn.
    M. IMPERFECTA Trin. Com. in unbroken soil below 3,000 feet. April—August.
DISTICHLIS Raf.
    D. MARITIMA Raf. Moist soil south of Pasadena. April—July.
DACTYLIS Linn.
    D. GLOMERATA Linn. Orchard Grass.) O. K., Santa Anita. June—August.
ACHYRODES Boehm.
    A. AUREUM (Linn.) O. Ktze. Com. below 2,000 feet. March—June.
POA Linn. Meadow grass.
    P. ANNUA Linn. (Goose grass.) Along streets and in lawns.
    P. PRATENSIS Linn. (Kentucky blue grass.) Along streets and in lawns. April—September.
    P. TENUIFOLIA Thurb. Oak Knoll. April—June.
ATROPIS Rupr.
    A. SCABRELLA Thurb. O. K., San G. Mts. April—June.
FESTUCA Linn.
    F. ELATIOR Linn. var. PRATENSIS Linn. O. K. April—June.
    F. MICROSTACHYS Nutt. O. K., A. S. April—July.
    F. MYURUS Linn. Com. below 3,000 feet. March—July.
    F. TENELLA Willd. Oak Knoll. April—June.
BROMUS Linn. Brome grass.
    B. CILIATUS Linn. O. K., S. G. Mts. April—August.
    B. HOOKERIANUS Thurb. Com. below 2,000 feet. April—July.

B. MAXIMUS Desf. Com. below 2,000 feet. April—July.
B. MOLLIS Linn. Along streets April—July.
B RIGIDUS Roth. Devil's Gate. April—July.

LOLIUM Linn. Darnel.
    L. PERENNE Linn. (Perennial rye grass) Com. in cultivated soil. April—July.
    L. TEMULENTUM Linn. (Bearded Darnel.) Com. in cultivated soil and along
    streets. April—July.

HORDEUM Linn Barley.
    H. MURINUM Linn. (Wild Barley) Com below 1,800 feet. March—July.

ELYMUS Linn. Wild Rye.
    E. AMERICANUS V. & S. Oak Knoll. April—June.
    E. CONDENSATUS Presl. O. K.; A. S.; foothills of S. G. Mts. April—July.
    E. SIBIRICUS Linn. S. G. Mts June—August.
    E. SITANION Schult. O. K. Along streets. April—June.
    E. TRITICOIDES Nutt. A. S.; foothills of S. G. Mts. June—September.

Cyperaceæ.

CYPERUS Linn.
    C. DIANDRUS Torr. Var. CAPITATUS Brit. Oak Knoll.
    C ERYTHRORRHIZOS Muhl. O. K.; Baldwin's Ranch. July—October.
    C. LAEVIGATUS Linn. Arroyo Seco. August—November.

SCIRPUS Linn. Bullrush. Club-rush.
    S. LACUSTRIS Linn. var. OCCIDENTALIS Wats. Oak Knoll. April—June.
    S. OLNEYI Gr. O. K ; A. S.
    S. RIPARIUS Spreng. Oak Knoll. February—June.
    S. SYLVATICUS Linn. var. MICROCARPUS (Presl.) McM. O. K.; A. S. May—
    August.

HELEOCHARIS R. Br. Spike-rush.
    H. ACICULARIS (Linn.) R Br. Oak Knoll.
    H. ARENICOLA Torr. Oak Knoll; A. S. April—June.
    H. PALUSTRIS (Linn.) R. Br.

CAREX Linn. Sedge.
    C. BARBARAE Dewey. O K.; R. C. February—June.
    C. FILIFORMIS Linn. var. LANUGINOSA (Michx.) B. P. S. O. K.; A. S. Feb-
    ruary—May.
    C. MULTICAULIS Bailey. Mt. Lowe. April—June.
    C. SPISSA Bailey. O. K.; R. C. February—May.
    C. TERETIUSCULA Good. var. RAMOSA Boott. R. C.; L. St. A. C. June—
    August.
    C. TRIQUETRA Bott. Hills near Arroyo Seco. March—May.

Lemnaceæ.

LEMNA Linn. Duckweed.
    L. GIBBA Linn. Johnson's Lake.
    L. VALDIVIANA Phil. O. K.; A. S.

Juncaceæ. Rush Family.

JUNCUS Linn. Bog-Rush.
    J. BALTICUS Deth. O. K ; A. S. May—July.
    J. BUFONIUS Linn. Com. in moist soil below 1000 feet. March—June.
    J. COMPRESSUS HBK. Oak Knoll. July—September.
    J. DUBIUS Eng. Oak Knoll. July—September.
    J. LONGISTYLIS Torr. O. K.; A. S. June—August.
    J. NODOSUS Linn. O. K.; A. S.; M. C. June—August.
    J. PHAEOCEPHALUS Eng. O. K.; A. S. June—August.
    J. ROBUSTUS Wats. O. K., A. S., M. C. April—August. Used by the Indians for
    basket making.
    J. XIPHIOIDES Mey. O. K., A. S. May—July.

43

Liliaceæ.   Lily Family.

ZYGADENUS Michx.
   Z. FREMONTI Torr.   Hills near Arroyo Seco.   March—April.
CALOCHORTUS Pursh.   Mariposa Lily.
   C. ALBUS Dougl.   A. S., San Gabriel Mts.   May—June.
   C. CATALINAE Wats.   Hills southwest of Pasadena.   February—May.
   C. SPLENDENS Dougl.   Hills near A. S., San Gabriel Mts.   May—June.
   C. WEEDII Wood var. PURPURASCENS Wats.   L. P., Hills near Pasadena.   June—
   July.
MUILLA Wats.
   M. SEROTINA Greene.   O. K., A. S, Echo Mt.   April—June.
LILIUM Linn.   Lily.
   L. HUMBOLTDII R. & L. (Wild Tiger Lily.)   A. S., canyons of San Gabriel Mts.
   June—July.
BRODLEA Smith.
   B. CAPITATA Benth. (Cluster Lily.)   Common below 3000 feet.   February—May.
   B. MINOR Wats.   South Los Robles Ave.   April—May.
BLOOMERIA Kell.
   B. AUREA (Hook.) Kell. (Golden Cluster Lily.)   O. K., A. S.   April—May.

YUCCA WHIPPLEI.· SPANISH BAYONET.
Showing one plant in bloom, and an old stem still retaining some of its seed pods.

CHLOROGALUM Kunth. Soap Plant.
    C. POMERIDIANUM Kunth. O. K., A. S. and adjacent hills, S. G. Mts. May and
    June.
FRITILLARIA Linn.
    F. BIFLORA Lindl. Arroy Seco. February—April.
YUCCA Linn. Spanish Bayonet.
    Y. WHIPPLEI Torr. Common on hills and in S. G. Mts. May—July.

Iridaceæ. Iris Family.

SISYRINCHIUM Linn. Blue-eyed Grass.
    S. BELLUM Wats. Common below 1,500 feet. February—April.

Orchidaceæ. Orchids.

EPIPACTIS Haller.
    E. GIGANTEA Dougl. A. S., Eaton C. April and May.
HABENARIA Willd.
    H. LEUCOSTACHYS Wats. Eaton Canyon. April—June.
    H. UNALASCHENSIS Wats. Hills near Pasadena, Echo Mt., W. C. April—June.

SUB-CLASS II. DICOTYLEDONES.

Division 1. Choripetalæ. Separate-petaled Plants.

Juglandaceæ.

JUGLANS Linn. Walnut.
    J. CALIFORNICA Wats. (California Walnut). Hills near Arroyo Seco. April.

Salicaceæ.

POPULUS Tourn. Poplar. Cottonwood.
    P. TRICHOCARPA Torr. & Gr. (Cottonwood.) A. S., M. C. March and April.
SALIX Tourn. Willow.
    S. LAEVIGATA Bebb. (Black willow.) O. K., A. S. April.
    S. LASIANDRA Benth. O. K., A. S. May.
    S. LASIOLEPIS Benth. (White Willow.) O. K., A. S , M. C., R. C. December—
    February.
    S. LONGIFOLIA Muhl. Arroyo Seco. March and April.

Betulaceæ.

ALNUS Tourn. Alder.
    A. RHOMBIFOLIA Nutt. A. S., Canyons of S. G. Mts. January and February.

Fagaceæ.

QUERCUS Linn. Oak.
    Q. AGRIFOLIA Nee. (Red Oak.) Com. from O. K. to Sierra Madre. April.
    Q. CHRYSOLEPIS Liebm. (Live Oak.) S. G. Mts. April.
    Q. DUMOSA Nutt. (Scrub Oak.) Hills near A. S., foothills of S. G. Mts. March
    and April.
    Q. ENGELMANNI Greene. (White Oak, Blue Oak.) Com. from O. K. to Sierra
    Madre.

Loranthaceæ.

PHORADENDRON Nutt. Mistletoe.
    P. FLAVESCENS (Pursh.) Nutt. On oaks, sycamore and alder at O. K. and in S. G.
    Mts. August.

Piperaceæ.

ANEMOPSIS Hook.
    A. CALIFORNICA Hook. (Yerba Mansa.) Oak Knoll. April—July.

Urticaceæ.

PARIETARIA Tourn.
  P. DEBILIS Forst. Canyons of S. G. Mts. April and May.
URTICA Linn. Nettle.
  U. HOLOSERICEA Nutt. Com. in shaded soil. March—May.
  U. URENS Linn. Com. in waste soil. April—June.

Platanaceæ.

PLATANUS Tourn. Sycamore.
  P. RACEMOSA Nutt. O. K., A. S., canyons of S. G. Mts. April.

Polygonaceæ.

RUMEX Linn. Dock. Sorrel
  R. ACETOSELLA Linn. (Sheep sorrel.) Along streets.
  R. CONGLOMERATUS Murr. (Green dock.) O. K., A S. April—June.
  R. CRISPUS Linn. (Curled dock.) Common in moist soil. February—June.
  R. HYMENOSEPALUS Torr. (Canaigre.) A. S., N. Pasadena. March and April.
  R. SALICIFOLIUS Wein. (White dock.) O. K., A. S. May and June.
POLYGONUM Linn. Knotweed.
  P. ACRE HBK. (Water smartweed.) O. K., A. S. April—July.
  P. AVICULARE Linn. (Knot-grass.) Com. in yards and along streets. May—July.
  P. CONVOLVULUS Linn. Fields. May—July.
  P. NODOSUM Pers. O. K., A. S., M. C., R. C. April—July.
ERIOGONUM Michx.
  E. ELONGATUM Benth. A. S., foothills of S. G. Mts. June—October.
  E. FASCICULATUM Benth. (Wild buckwheat.) Com. in dry unbroken soil. May—September.
  E. GRACILE Benth. Arroyo Seco. April—June.
  E. SAXATILE Wats. Com. above 4,000 feet in S. G. Mts. June—August.
  E. VIRGATUM Benth. Common below 3,000 feet. June—September.
CHORIZANTHE R. Brown.
  C. PROCUMBENS Nutt. Lincoln Park. May—July.
  C. STATICOIDES Benth. Common below 3,000 feet. May—August.
PTEROSTEGIA Fisch. & Mey.
  P. DRYMARIOIDES F. & M. Com. in unbroken soil below 4,000 feet. March—June

Chenopodiaceæ

CHENOPODIUM Tour. Lamb's quarter. Goosefoot.
  C. ALBUM Linn. (Pigweed.) Common in cultivated soil. March—September.
  C. AMBROSIOIDES Linn. (Mexican Tea.) Com. in waste soil. March—October.
  C. MURALE Linn. Com. below 1,800 feet, especially in cultivated soil. February—June.
ATRIPLEX Tourn.
  A. MICROCARPA Dietr. Along streets. July—October.
  A. PATULA Linn. Along streets. August—October.

Amarantaceæ.

AMARANTUS Tourn.
  A. ALBUS Linn. (Tumbleweed.) Waste ground. May—September.
  A. RETROFLEXUS Linn. Cultivated soil. May—September.

Nyctoginaceæ.

MIRABILIS Linn. Four o'clock.
  M. CALIFORNICA Gr. Common on dry hills below 4,000 feet. February—July.

Portulacaceæ.

CALYPTRIDIUM Nutt.
  C. MONANDRUM Nutt. Hills near Arroyo Seco. March—June.

CLAYTONIA Linn.
  C. PERFOLIATA Donn. (Indian lettuce.) Com. in shaded soil below 4,000 feet.
  February—May.
CALANDRINIA HBK.
  C. MENZIESII Hook. Com. below 2,000 feet. January—April.
PORTULACA Tourn.
  P. OLERACEA Linn. (Purslane.) Com. in cultivated and waste ground. June—
  September.

Caryophyllaceæ.

SILENE Linn. Catchfly.
  S. ANTIRRHINA Linn. Frequent below 2,000 feet. April—June.
  S. GALICA Linn. Common below 2,000 feet.
  S. LACINIATA Cav. (Indian pink.) Abundant on dry hills below 3,000 feet. May
  —July.
  S. PALMERI Wats. S. G. Mts. above 4,500 feet. June—July.
  S. PLATYOTA Wats. S. G. Mts. above 3,500 feet. June—July.
STELLARIA Linn.
  S. MEDIA Linn. (Chickweed.) Com. in shaded soil below 1,500 feet.
  S. NITENS Nutt. O. K., A. S., Canyons of S. G. Mts. April—June.
CERASTIUM Linn. Mouse-ear.
  C. VULGATUM Linn. Com. in lawns and waste soil.
ARENARIA Linn.
  A. DOUGLASII (Fen.) T. & G. Frequent in dry soil below 2,000 feet. April—May.
SAGINA Linn.
  S. OCCIDENTALIS Wats. Moist and shaded soil about Pasadena. April.
SPERGULA Linn. Corn-spurrey.
  S. ARVENSIS Linn. Frequent along streets and in yards. March—May.
POLYCARPON Linn.
  P. DEPRESSUM Nutt. Altadena. April—June.

Ceratophyllaceæ.

CERATOPHYLLUM Linn. Hornwort.
  C. DEMERSUM Linn. In ponds at Oak Knoll. June—August.

Ranunculaceæ.

AQUILEGIA Linn. Columbine.
  A. TRUNCATA F. & M. Oak Knoll. May—August.
DELPHINIUM Tourn. Larkspur.
  D. CARDINALE Hook. (Scarlet larkspur.) Dry hillsides below 3,000 feet. June—
  August.
  D. DECORUM F. & M. Hillsides. May—July.
  D. PARRYI Gr. Foothills S. G Mts. April—June.
  D. VARIEGATUM T. & G. Hillsides. May—July.
PÆONIA Linn. Pæony.
  P. CALIFORNICA Nutt. (California Pæony.) Com. in unbroken soil below 2,000
  feet. March—May.
CLEMATIS Linn. Virgin's Bower.
  C. LIGUSTICIFOLIA Nutt. O. K., A. S., S. G. Mts. below 4,000 feet.
RANUNCULUS Linn. Buttercup.
  R. CALIFORNICUS Benth. O. K., A. S. February—May.
  R. HEBECARPUS H. & A. Oak Knoll. March—May.
THALICTRUM Tourn. Meadow Rue.
  T. POLYCARPUM Wats. O. K., A. S., Canyons of S. G. Mts. May—July.

Laurineæ.

UMBELLULARIA Nutt. Mountain Laurel. Spice-tree. Bay-tree.
  U. CALIFORNICA (H. & A.) Nutt. Canyons of S. G. Mts. February—April.

44

Papaveraceæ.

PAPAVER Tourn. Poppy.
  P. CALIFORNICUM Gr. Hills near Pasadena. April—June.
ARGEMONE Linn.
  A. HISPIDA Gr. Arroyo Seco. April—June.
PLATYSTEMON Benth.
  P. CALIFORNICUS Benth. (Cream-cups.) Com. below 2,000 feet. March—May.
  P. DENTICULATUS Greene. A. S., Canyons of S. G. Mts. April—May.
DENDROMECON Benth.
  D. RIGIDUM Benth. (Tree poppy.) Hills near Arroy Seco. March—June.

From " Land of Sunshine "
June 1894.

ESCHSCHOLTZIA CALIFORNICA.—CALIFORNIA POPPY.
One-third natural size.

ESCHSCHOLTZIA Cham.
  E. CALIFORNICA Cham. (California poppy.) Com. below 2,000 feet. Most abundant from January to May, but to be found at all times of the year.

Fumariaceæ.

DICENTRA Bork. Ear-drop.
  D. CHRYSANTHA H. & A. Frequent in dry soil, especially in S. G. Mts. May—July.

Cruciferæ.

TROPIDOCARPUM Hook.
  T. GRACILE Hook. Oak Knoll. March—May.
RAPHANUS Linn.
  R. SATIVUS Linn. (Radish.) Com. in fields. April—June.
THYSANOCARPUS Hook. Lace-pod.
  T. CURVIPES Hook. O. K., A. S., Canyons of S. G. Mts. February—April.
  T. LACINIATUS Nutt. O. K., A. S., Canyons of S. G. Mts. February—April.
  T. PUSILLUS Hook. O. K., A. S. February—April.

LEPIDIUM Linn. Pepper-grass.
    L. INTERMEDIUM Gr. Com. below 1800 feet. April—June.
    L. NITIDUM Nutt. Com. below 1800 feet. February—April.
BURSA Sieg.
    B. PASTORIS (Linn.) Wigg. Com. below 1800 feet. March—June.
SISYMBRIUM Linn.
    S. CANESCENS Nutt. O. K., A. S., plains adjacent to S. G. Mts. February—April.
BRASSICA Linn. Mustard.
    B. CAMPESTRIS Linn. Com below 1500 feet. February—April.
    B NIGRA (Linn.) Kochz. Com. below 2000 feet. April—June.
ERYSIMUM Linn.
    E. ASPERUM (Nutt.) DC. A. S., canyons of San Gabriel Mts. February—April.
NASTURTIUM R. Br.
    N. OFFICINALE R. Br. (Water-cress) Abundant in A. S. and at O. K.
CARDAMINE Linn.
    C. GAMBELII Wats. Oak Knoll. May—July.
    C. INTEGRIFOLIA (Nutt.) Greene. O. K., A. S., M. C February—April.
STREPHANTHUS Nutt.
    S. HETEROPHYLLUS Nutt. Hills near Arroyo Seco. March—April.
ARABIS Linn.
    A. GLABRA (Linn.) Wein. Hills near Arroyo Seco. March—April.
    A. HOLBOELLII Hornem Com. in San Gabriel Mts. April—June.
ALYSSUM Linn.
    A. MARITIMUM (Linn.) L. Am. Com. along streets.

Violaceæ.

VIOLA Linn. Violet.
    V. PEDUNCULATA T. & G. O. K., A. S., Las Casitas. February—April.

Cistaceæ.

HELIANTHEMUM Tourn. Rock-Rose.
    H. SCOPARIUM Nutt. Common on dry hills below 4000 feet.—April—June.

Lythraceæ.

LYTHRUM Linn.
    L. CALIFORNICUM Torr & Gr. O. K., A. S. April—June.

Malvaceæ.

MALVA Linn. Mallow.
    M. PARVIFLORA Linn. Common below 2000 feet. February—April.
SIDALCEA Gray.
    S. DELPHINIFOLIA (Nutt.) Green. Oak Knoll. February—May

Geraniaceæ.

GERANIUM Linn. Cranesbill.
    G. CAROLINIANUM Linn (Wild Geranium.) O. K., A. S. March—May.
ERODIUM L'Her. Storksbill.
    E. CICUTARIUM (Linn.) L'Her. (Alfilaria) Common below 2000 feet. January—June.
    E. MOSCHATUM (Linn.) L'Her. (Alfilaria) Com. below 2000 feet. January—June.
OXALIS Linn. Wood-Sorrel.
    O. CORNICULATA Linn. Frequent along streets.

Anacardiaceæ.

RHUS Linn. Sumach.
    R. DIVERSILOBA T. & G. (Poison sumach. "Poison Oak") O. K., A. S., S. G. Mts. April—May.
    R. INTEGRIFOLIA B. & H. Com. in unbroken soil below 4000 feet. March—May.

R. LAURINA Nutt. Com. in unbroken soil below 4000 feet. September—October. Berries used by the Indians for making an acid drink.
R. TRILOBATA Nutt. O. K., A. S., canyons of S. G. Mts. February—April. Long. slender branches used by Indians for making baskets.

Sapindaceæ.

ACER Tourn. Maple.
A. MACROPHYLLUM Pursh. Canyons of S. G. Mts. March—April.

Polygalaceæ.

POLYGALA Tourn
P. CALIFORNICA Nutt. Near old Wilson trail. June—August.

Ampilidaceæ.

VITIS Tourn. Grape.
V. CALIFORNICA Benth. O. K., A. S., canyons of S. G. Mts. May—July.

Rhamnaceæ.

RHAMNUS Linn. Buckthorn.
R. CALIFORNICA Esch. O. K., A. S., canyons of S. G. Mts. February—April.
R. CROCEA Nutt. Arroyo Seco, S. G. Mts. below 5000 feet. February—March. Berries said to be used by the Indians for food.
CEANOTHUS Linn. California Lilac. Buckthorn.
C. CRASSIFOLIUS Torr. A. S., canyons of S. G. Mts. February—April.
C. CUNEATUS Nutt. Near Sierra Madre. March—May.
C. DIVARICATUS Nutt. Abundant in S. G. Mts. 2000 feet, 5000 feet. March — April.
C. INTEGERRIMUS H. & A. S. G. Mts. April—May.
C. OGLIGANTHUS Nutt. S. G. Mts. March—April.

Euphorbiaceæ.

CROTON Linn.
C. CALIFORNICUS Muell. From center of Pasadena to foot of S. G. Mts. April—June.
C. SETIGERUS Hook. Com. below 2000 feet. August—November.
EUPHORBIA Linn.
E. ALBOMARGINATA T. & G. Com. below 2000 feet. April—July.
E. POLYCARPA Benth. var. VESTITA Wats. Foothills S. G. Mts.
E. SERPYLLIFOLIA Pers. Frequent below 2000 feet. May—August.

Umbelliferæ

HYDROCOTYLE Tourn.
H. RANUNCULOIDES Linn. Oak Knoll. May—July.
SANICULA Tourn.
S. BIPINNATA H. & A. O. K., A. S. March—May.
S. BIPINNATIFIDA Dongl. Com. in open ground below 2,000 feet. March—May.
S. MENZIESII H. & A. Com. in shaded soil below 3,000 feet. March—May.
S. NUDICAULIS H. & A. Com. in open ground below 2,000 feet. March—May.
S. TUBEROSA Torr. O. K., A. S. April and May.
VELAEA DC.
V. ARGUTA T. & G. C. & R. Hills near A. S., S. G. Mts. March—May.
V. PARISHII C. & R. Mount Lowe. June and July.
CONIUM Linn.
C. MACULATUM Linn. Oak Knoll. March—June.
SIUM Linn.
S. CICUTÆFOLIUM Gmel. Oak Knoll. June—August.
OSMORRHIZA Raf.
O. BRACHYPODA Torr. O. K., A. S., canyons of S. G. Mts. March—May.
ŒNANTHE Linn.
Œ. CALIFORNICA Wats. Oak Knoll. June—August.

APIUM Linn. Celery.
    A. GRAVEOLENS Linn. (Wild Celery.) O. K., A. S. March-August.
APIASTRUM Autt.
    A. ANGUSTIFOLIUM Nutt. Common below 2,000 feet. April—June.
FOENICULUM Adans. Fennel.
    F. VULGARE Ger. O. K. Along streets. May—July.
PEUCEDANUM Linn.
    P. CARUIFOLIUM T. & G. Hills near Arroyo Seco. April—June.
    P. HASSEI C. & R. Hills near Arroyo Seco. March—May.
    P. UTRICULATUM Nutt. Com. in unbroken soil below 1,500 feet. March—May.
DAUCUS Tourn. Carrot.
    D. PUSILLUS Michx. (Wild carrot) Com. below 1,800 feet. March—May.
CAUCALIS Linn.
    C. MICROCARPA H. & A. Oak Knoll. April and May.
    C. NODOSA Hudson. Oak Knoll. April and May.

Araliaceæ.

ARALIA Linn.
    A. CALIFORNICA Wats. (Spikenard.) Canyons of S. G. Mts. June—August.

Cornaceæ.

CORNUS Linn. Cornel. Dogwood.
    C. OCCIDENTALIS (T. & G.) Cov. Oak Knoll. April—September.
GARRYA Dougl.
    G. VEATCHII Kell. Echo Mt. May—July.

Saxifragaceæ.

SAXIFRAGA Linn. Saxifrage.
    S. CALIFORNICA Greene. Arroyo Seco. April—June.
BOYKINIA Nutt.
    B. ROTUNDIFOLIA Parry. Canyons of S. G. Mts. June—August.
TELLIMA R. Br.
    T. AFFINIS (Gray) Bol. O. K., A. S., canyons of S. G. Mts. March and April.
HEUCHERA Linn.
    H. RUBESCENS Torr. Frequent in S. G. Mts., above 5,000. June—August.
RIBES Linn. Currant. Gooseberry.
    R. AMARUM McC. Erythea II, 79. Canyons of S. G. Mts. February—March.
    R. DIVARICATUM Dougl. Oak Knoll. February—May.
    R. GLUTINOSUM Benth. S. G. Mts. 2,000—5,000 feet. January—March.
    R. HESPERIUM McC. Erythea. II, 79. Canyons of S. G. Mts. January and
    February.
    R. SPECIOSUM Pursh. Hills near Arroyo Seco. January—March.
    R. TENUIFLORUM Lindl. Hills near Arroyo Seco. February and March.

Crassulaceæ.

TILLAEA Linn.
    T. MINIMA Miers. Common below 3,000 feet February—May.
COTYLEDON Linn.
    C. LANCEOLATA Benth. Oak Knoll, foothills of S. G. Mts. June and July.

Cactaceæ.

CEREUS Haw. Cactus.
    C. EMORYI Engelm. Arroy Seco. May and June.
OPUNTIA Tourn.
    O. ENGELMANNI Salm. (Prickly Pear.) Com. in dry unbroken soil. May and
    June.

45

Loasaceæ.

MENTZELIA Linn.
    M. ALBICAULIS Dougl. Near Sierra Madre. July—September.
    M. MICRANTHA T. & G. Foothills S. G. Mts. May—July.

Datisaceæ.

DATISCA Linn.
    D. GLOMERATA (Presl.) Brew. & Wat. O. K., A. S., canyons S. G. Mts. May—July.

Onagraceæ.

EPILOBIUM Linn.
    E. ADENOCAULON Haussk. var. OCCIDENTALE Trel. Com. in wet soil. April—August.
    E. HOLOSERICEUM Trel. Oak Knoll. May—August.
    E. PANICULATUM Nutt. Arroyo Seco. August and September.
ZAUSCHNERIA Presl. "Wild Fuchsia."
    Z. CALIFORNICA Presl. S. G. Mts. 3,000—5,000 feet.
    Z. CALIFORNICA Presl. var. MICROPHYLLA Gray. Com. on dry hills.
EULOBUS Nutt.
    E. CALIFORNICUS Nutt. Frequent below 3,000 feet. May—July.
ŒNOTHERA Linn. Evening Primrose.
    Œ. ALYSSOIDES H. & A. Hills near Pasadena, Echo Mt. May—July.
    Œ. BIENNIS L. Common in canyons. May—September.
    Œ. BISTORTA Nutt. Com. below 2,000 feet. February—June.
    Œ. MICRANTHA Horn. Frequent below 2,000 feet. March—June.
    Œ. STRIGULOSA T. & G. Frequent below 2,000 feet March—June.
GODETIA Spach.
    G. BOTTÆ Spach. Hills near Arroyo Seco, S. G. Mts. April—July.
    G. EPILOBIOIDES (T. & G.) Wats. Hills near Arroyo Seco. April—May.
    G. QUADRIVULNERA (Dougl.) Wats. Hills near A. S., foothills S. G. Mts. April—July.
JUSSLÆA Linn.
    J. DIFFUSA Forsk. Shorb's ranch. May—August.

Rosaceæ.

PRUNUS Tourn
    P. ILICIFOLIA (Nutt.) Walp. (Wild cherry.) Foothills S. G. Mts. May.
HETEROMELES Roemer.
    H. ARBUTIFOLIA (Ait.) Rœmer. (California Holly.) Hills along Arroyo Seco, foothills S. G. Mts. June—August.
HOLODISCUS Maxim. Meadow-Sweet.
    H. DISCOLOR (Pursh.) Max. S. G. Mts. May—August.
ADENOSTOMA Hook. & Arn.
    A. FASCICULATUM H. & A. Com. in dry soil below 4,000 feet. May—June. Sold by Mexicans for fuel, as "Grease-wood."
CERCOCARPUS HBK. Mountain Mahogany.
    C. PARVIFOLIUS Nutt. Foothills S. G. Mts. March—April.
POTENTILLA Linn. Five-finger.
    P. CALIFORNICA (C. & S.) Greene. Frequent below 2,500 feet. March—May.
    P. GLANDULOSA Lindl. Frequent below 2,500 feet. April—June.
ALCHEMILLA Tourn.
    A. ARVENSIS Scop. Oak Knoll. March—May.
RUBUS Linn. Blackberry. Raspberry.
    R. URSINUS C. & S. (Wild blackberry.) Com. in shaded soil. March—May.
ROSA Tourn. Rose.
    R. CALIFORNICA C. & S. O. K., A. S., Canyons S. G. Mts.

VICIA Tourn. Vetch.
    V. AMERICANA Muhl. O. K., A. S. March—May.
    V. EXIGUA Nutt. Arroyo Seco. March—April.
    V. SATIVA Linn. O. K., Sierra Madre. March—April.
LATHYRUS Linn.
    L. VESTITUS Nutt. (Wild Pea.) A. S., O. K., Canyons S. G. Mts. February—
    May.
ASTRAGALUS Tourn. Rattle-Weed.
    A. LEUCOPSIS Torr. Com. in open unbroken soil. March—May.
AMORPHA Linn.
    A. CALIFORNICA Nutt. S. G. Mts. 3,000 feet. May—June.
PSORALEA Linn.
    P. MACROSTACHYS DC. A S., Canyons south of Pasadena. May—June.
    P. ORBICULARIS Lindl. Arroyo Seco. April—May.
    P. PHYSODES Dougl. Arroyo Seco. June—July.
LOTUS Tourn.
    L. AMERICANUS (Nutt.) Bisch. Frequent below 1,500 feet. May—July.
    L. ARGOPHYLLUS (Gray) Greene. S. G. Mts. 3,000–5,000 feet. April—June.
    L. GLABER (Torr.) Greene. Com. in unbroken soil below 3,000 feet.
    L. NEVADENSIS (Wats.) Greene. A. S., Canyons of S. G. Mts. below 2,500 feet.
    April—July.
    L. OBLONGIFOLIUS (Benth.) Greene. Arroyo Seco. April—June.
    L. SALSUGINOSUS Greene. Frequent in shaded soil below 1,500 feet. March—
    April.
    L. STRIGOSUS (Nutt.) Greene. Com. below 2,000 feet. January—May.
TRIFOLIUM Linn. Clover.
    T. CILIOLATUM Benth. Com. below 1,800 feet. April—May.
    T. GRACILENTUM T. & G. Com. below 1,800 feet. April—May.
    T. INVOLUCRATUM Willd. Frequent below 1,800 feet April—May.
    T. MACRÆI H. & A. var. ALBOPURPUREUM (T. & G.) Greene. Com. below 1,800 feet.
    April—May.
    T. MICROCEPHALUM Pursh. Com. below 1,800 feet. April—May.
    T. ROSCIDUM Greene. O. K., A. S., Canyons of S. G. Mts. below 2,500 feet. May—
    August.
    T. STENOPHYLLUM Nutt. Oak Knoll. March—April.
MELILOTUS Tourn. Sweet clover.
    M. ALBA Lam. Arroyo Seco. May—August.
    M. INDICA All. Com. in shaded soil below 1,500 feet. April—July.
MEDICAGO Linn.
    M. DENTICULATA Willd. (Burr-clover.) Com. below 2,000 feet. January—June.
    M. SATIVA Linn. (Alfalfa.) Frequent along streets and roads whither it has es-
    caped from fields.
LUPINUS Linn. Lupine.
    L. AFFINIS Ag. Oak Knoll. March—May.
    L. ALBIFRONS Benth. Arroyo Seco. April—June.
    L. CYTISOIDES Ag. A. S., Canyons of S. G. Mts. below 3,000 feet. May—July.
    L. FORMOSUS Greene var. BRIDGESII (Wats.) Greene. Com. below 2,500.
    L. HIRSUTISSIMUS Benth. O. K., A. S., foothills S. G. Mts. March—May.
    L. MICRANTHUS Dougl. Com. below 2,000 feet. January—May.
    L. SPARSIFLORUS Benth. Frequent below 3,000 feet. March—May.
    L. TRUNCATUS Nutt. Com. below 3,000 feet. March—May.

DIVISION II. SYMPATALÆ United-petaled Plants.

Ericaceæ.

ARBUTUS Tourn.
    A. MENZIESII Pursh. (Madroña.) S. G. Mts. Confined to a narrow belt at about
    3,000 feet. March—April.
ARCTOSTAPHYLOS Adans. Manzanita.
    A. PUNGENS HBK. S. G. Mts. 4,000 feet to near summit. February—March.
    A. TOMENTOSA Dougl. S. G. Mts. 2,000–4,500 feet. March—April.

Primulaceæ.

DODECATHEON Linn.
  D. CLEVELANDI Greene. (Shooting Star.) Frequent below 2,000 feet. February—April.
ANAGALLIS Tourn. Pimpernel.
  A. ARVENSIS Linn. Common below 1,800 feet.
SAMOLUS Linn.
  S. VALERANDI Linn. var. AMERICANA Gr. O. K., A. S., canyons of San Gabriel Mts. June—November.

Convolvulaceæ.

CUSCUTA Tourn. Dodder.
  C. SUBINCLUSA Dur. & Hilg. Common on shrubs and other plants. May—July.
CONVOLVULUS Linn. Bindweed.
  C. OCCIDENTALIS Gray. (Wild morning-glory) Common below 2,500 feet. February—July.

Polemoniaceæ.

NAVERRETIA Ruiz & Pav.
  N. ATRACTYLOIDES (Benth.) H. & A. Common in dry soil below 2,500 feet. May—July.
  N. PROSTRATA (Gr.) Greene. Oak Knoll. April—May.
  N. VISCIDULA Benth. Common in dry soil below 2,500 feet. June—August.
LEPTODACTYLON H. & A.
  L. CALIFORNICUM H. & A. Common on dry soil below 4,000 feet. April—June.
GILIA Ruiz & Pav.
  G. ACHILLEÆFOLIA Benth. Frequent below 2,000 feet. April—June.
  G. GILIOIDES (Benth.) Greene. A. S. and adjacent hills. April—July.
  G. INCONSPICUA Dougl. Frequent below 2,000 feet. April—June.
  G. MULTICAULIS Benth. Common below 2,000 feet. April—June.
  G. TENUIFLORA Benth. Frequent in S. G. Mts. June—August.
  G. VIRGATA Steud. Arroyo Seco. May—August.
LINANTHUS Benth.
  L. ANDROSACEUS (Benth.) Greene. Frequent below 2,000 feet. April—June.
  L. DIANTHIFLORUS (Benth.) Greene. Common below 2,000 feet. February—May.
  L. PHARNACEOIDES (Benth.) Greene. Common in dry soil below 2,000 feet. May—July.
PHLOX Linn.
  P. GRACILIS (Dougl.) Greene. Arroyo Seco. February—April.

Hydrophyllaceæ.

NEMOPHILA Nutt.
  N. AURITA Lindl. Arroyo Seco. April—June.
  N. INSIGNIS Dougl. Com. in shaded soil below 3,000 feet. February—May.
  N. MENZIESII H. & A. Com. in shaded soil below 3,000 feet. February—May.
EUCRYPTA Nutt.
  E. CHRYSANTHEMIFOLIA (Benth.) Greene. Com. in shaded soil below 3,000 feet. March—June.
PHACELIA Juss.
  P. BRACHYLOBA Gray. S. G. Mts. 3,000-5,000 feet. June—August.
  P. CIRCINATA (Willd.) Jaq. Common below 2,500 feet. May—July.
  P. HISPIDA Gr. Common below 2,500 feet. March—July.
  P. LONGIPES Torr. A. S., S. G. Mts. May—July.
  P. RAMOSISSIMA Hook. O. K., A. S., canyons of S. G. Mts. April—July.
  P. TANACETIFOLIA Benth. Common below 2,500 feet. March—June.
  P. WHITLAVIA Gray. Common on dry hillsides below 3,000 feet. March—May.
EMMENANTHE Benth.
  E. PENDULIFLORA Benth. Frequent below 2,500 feet. April—June.

ERIODICTYON Benth.
  E. CALIFORNICUM (H. & A.) Greene. (Verba Santa.) S. G. Mts. 2,000-4,000 feet.
  May—July. Leaves used by the Mexicans as a tonic.
  E. TOMENTOSUM Benth. S. G. Mts. 2,000-4,000 feet. May—July.

Borraginaceæ.

PECTOCARYA DC.
  P. LINEARIS (R. & P.) DC. O. K., A. S. March—May.
PLAGIOBOTHRYS Fisch. & Mey.
  P. NOTHOFULVUS Gr. Common below 2,000 feet. February—May.
CRYPTANTHE Lehm.
  C. AMBIGUA (Gr.) Greene. Common below 2,000 feet. April—June.
  C. MICROSTACHYS Greene. Common below 2,000 feet. April—June.
AMSINCKIA Lehm.
  A. SPECTABILS F. & M. Com. below 2,000 feet. March—May.

Solanaceæ.

SOLANUM Tourn.
  S. DOUGLASII Dunal. Com. below 2,500 feet.
  S. UMBELLIFERUM Esch. S. G. Mts. April—June.
  S. XANTI Gray S. G. Mts. April—June.
PHYSALIS Linn.
  P. AEQUATA Jacq. Com. in cultivated and waste ground. May—September.
DATURA Linn.
  D. METELOIDES DC. In moist and shaded soil below 1,000 feet. May—October.

Labiatæ.

TRICHOSTEMA Linn. Blue Curls.
  T. LANATUM Benth. Echo Mt. May—July.
  T. LANCEOLATUM Benth. Com. below 2,000 feet. August—October.
MENTHA Linn. Mint.
  M. CANADENSIS Linn. Canyons south of Pasadena. May—September.
PYCNANTHEMUM Michx. Mountain Mint.
  P. CALIFORNICUM Torr. O. K., A. S., canyons of S. G. Mts. June—September.
MONARDELLA Benth.
  M. LANCEOLATA Gray. A. S., N. Pas., Las Casitas. June—September.
SALVIA Linn. Sage.
  S. CARDUACEA Benth. Occasional in dry soil. April—June.
  S. COLUMBARIAE Benth. (Chia.) Com. below 3,000 feet. March—May. Seeds
  used by Indians for food.
  S. MELLIFERA Greene. (Black Sage.) Com. on dry hillsides. April—June.
  S. SPATHACEA Greene. On banks of A. S. April—July.
RAMONA Greene.
  R. POLYSTACHYA (Benth.) Greene. (White Sage.) Com. on dry hillsides. April
  —June.
SCUTELLARIA Linn. Skull-cap.
  S. TUBEROSA Benth. Com. in unbroken soil below 1,500 feet. March—June.
MARRUBIUM Linn. Hoarhound.
  M. VULGARE Linn. Com. below 1,800 feet. March—July.
STACHYS Linn.
  S. ALBENS Gr. O. K., A. S., canyons of S. G. Mts. April—June.
  S. CALIFORNICA Benth. O. K., A. S., canyons of S. G. Mts. May—July.

Scrophulariaceæ.

CORDYLANTHUS Nutt.
  C. FILIFOLIUS Nutt. A. S., S. G. Mts. below 3,000 feet. June—September.

CASTILLEIA Linn. Painted-cup.
C. FOLIOLOSA H. & A. Com. on dry hillsides below 3,000 feet. April—June.
C. MINIATA Dougl. Com. on hillsides below 2,500 feet. April—June.
C. PARVIFLORA Bong. Hills near Arroyo Seco, R. C. March—May.
C. STENANTHA Gray. O. K., M. C. May—August.

ORTHOCARPUS Nutt.
O. PURPURASCENS Benth. ("Paint-brush.") Com. below 2,000 feet. March—May.

VERONICA Linn. Speedwell.
V. PEREGRINA Linn. Oak Knoll. May—June.

MIMULUS Linn. Monkey-flower.
M. BREVIPES Benth. Frequent below 3,000 feet. April—June.
M. CARDINALIS Dougl. O. K., A. S., canyons of S. G. Mts. June—October.
M. FLORIBUNDUS Gray. Arroyo Seco, canyons of S. G. Mts. June—October.
M. FREMONTI Gray. Echo Mt. April—June.
M. LUTEUS Linn. O. K., A. S.
M. LUTEUS Linn. var. DEPAUPERATUS Gr. S. G. Mts. 4,000-5,000. May—July.

DIPLACUS Nutt.
D. GLUTINOSUS (Wendl.) Nutt. Hills along Arroyo Seco. April—July.

PENTSTEMON Mitch. Beard-tongue.
P. AZUREUS Benth. Frequent on hillsides below 3,000 feet. May—July.
P. CENTRANTHIFOLIUS Benth. S. G. Mts. 2,000-5,000 feet. March—June.
P. CORDIFOLIUS Benth. Hills along A. S., S. G. Mts. below 3,000 feet. May—July.
P. PALMERI Gr. Mt. Lowe. June—August.
P. SPECTABILIS Thurb. Near Sierra Madre. April—June.
P. TERNATUS Torr. S. G. Mts. 4,000-5,000 feet. June—August.

COLLINSIA Nutt.
C. BICOLOR Benth. Com. below 3,000 feet. April—June.
C. PARRYI Gray. Arroyo Seco. April—May.

SCROPHULARIA Tourn. Figwort.
S. CALIFORNICA Cham. O. K.; A. S., canyons of S. G. Mts. May—August.

ANTIRRHINUM Tourn. Snapdragon.
A. COULTERIANUM Benth. Hills along A. S., foothills S. G. Mts. May—July.
A. GLANDULOSUM Lindl. A. S., Altadena. April—June.
A. NUTTALLIANUM Benth. Hills along A. S. May—June.
A. STRICTUM (H. & A.) Gr. Hills along A. S., R. C. May—July.

LINARIA Tourn. Toad-flax.
L. CANADENSIS (Linn.) Dum. Common below 2,000 feet. April—September.

VERBASCUM Linn. Mullein.
V. VIRGATUM With. Frequent below 2,000 feet, abundant at Altadena. June—August.

Orobanchaceæ.

APHYLLON Mitch.
A. TUBEROSUM Gr. Echo Mt. among shrubs. June—August.

Verbenaceæ.

VERBENA Linn.
V. POLYSTACHYA HBK. O. K., A. S. May—September.
V. PROSTRATA R. Br. Com. below 1,500 feet. April—September.

Plantaginaceæ.

PLANTAGO Linn.
P. HIRTELLA HBK. Oak Knoll. May—July.
P. LANCEOLATA Linn. Com. below 1,500 feet. April—July.
P. MAJOR Linn. Frequent below 1,500 feet. May—August.
P. PATAGONICA Jacq. Com. in dry soil below 2,500 feet. March—May.

Gentanaceæ.

ERYTHRÆA Rn. Canchalagua.
E. VENUSTA Gr. Common in unbroken soil below 3,000 feet. April—July. In general use among the Mexicans as a medicinal herb.

Apocynaceæ.

APOCYNUM Tourn.
    A. CANNABINUM Linn. Indian Hemp.) In A. S. April—June.

Asclepiadaceæ.

ASCLEPIAS Linn. Milkweed.
    A. ERIOCARPA Benth. Frequent below 2,000 feet. Com. at Las Casitas. June—August.
    A. FASCICULARIS Decsne. Frequent below 2,000 feet. June—August.

Campanulaceæ

SPECULARIA Heist.
    S. BIFLORA Gray. Occasional below 1,500 feet. April—June.

Lobeliaceæ.

NEMACLADUS Nutt.
    N. RAMOSISSIMUS Nutt. var. PINNATIFIDUS. (Greene) Gray. Echo Mt. May—July.
LOBELIA Linn.
    L. SPLENDENS Willd. Oak Knoll. Wild-grape canyon. July—October.
PALMERELLA Gray.
    P. DEBILIS Gr. var. SERRATA Gr. A. S., canyons S. G. Mts. July—October.

Cucurbitaceæ.

CUCURBITA Linn. Cucumber. Squash. Pumpkin.
    C. FOETIDISSIMA HBK. (Mock Orange.) Common in dry soil below 2,000 feet. April—June.
MICRAMPELIS Raf. Big-root.
    M. FABACEA (Naud) Greene. O. K., A. S., foothills S. G. Mts. February—April.

Rubiaceæ.

GALIUM. Linn. Bedstraw.
    G. ANGUSTIFOLIUM Nut. Com. below 3,000 feet. March—June.
    G. APARINE Linn. O. K., A. S., canyons S. G. Mts. March—May.
    G. CALIFORNICUM Hook & Arn. O. K., A. S., foothills S. G. Mts. April—June.
    G. GRANDE McC. Erythea II: 124. S. G. Mts. 2,500—4,000 feet. April—June.
    G. NUTALLII Gr. Foothills S. G. Mts. May and June.
    G. OCCIDENTALIS McC. Erythea II: 124. S. G. Mts. 3,000—4,000 feet. May and June.
    G. TRIFIDUM Linn. Baldwin's Ranch. June and July.

Caprifoliaceæ.

SAMBUCUS Tourn. Elder.
    S. GLAUCA Nutt. Frequent below 2,500. April and May.
SYMPHORICARPUS Dill.
    S. MOLLIS Nutt. O. K., A. S., canyons S. G. Mts. April and May.
CAPRIFOLIUM DC.
    C. SUBSPICATA (H. & A.) Greene.

Dipsaceæ

DIPSACUS Tourn. Teasel.
    D. FULLONUM Linn. (Fuller's Teasel.) O. K. June—August.
SCABIOSA Linn.
    S. STELLATA Linn. Along streets in Altadena. May—July.

COLEOSANTHUS Cass.
    C. CALIFORNICUS (T. & G.) O. Ktze. Hills along Arroyo Seco, foothills S. G. Mts.
        July—September.
    C. NEVINII (Gr.) O. Ktze. Foothills S. G. Mts. July—September.

GRINDELIA Willd.
    G. ROBUSTA Nutt. Frequent below 2,000 feet. May—July.

HETEROTHECA Cass.
    H. GRANDIFLORA Nutt. Common below 2,000 feet.

CHRYSOPSIS Nutt.
    C. SESSILIFLORA Nutt. O. K., A. S. June—August.

ERICAMERIA Nutt.
    E. CUNEATUS (Gr.) McC. Rubio Canyon. June—September.
    E. MONACTIS (Gr.) McC. Hills along A. S., foothills S. G. Mts. June—Sept.

HAZARDIA Greene.
    H. SQUARROSUS (H. & A.) Greene. Hills along A. S., foothills S. G. Mts. June—
        September.

BIGELOVIA DC.
    B. VENETA Gr. A. S., foothills S. G. Mts. and adjacent plains. July—Sept.

SOLIDAGO Linn.
    S. CALIFORNICA Nutt. O. K., A. S., canyons of S. G. Mts. to 5,000 feet (side of
        Mt. Lowe). August—October.
    S. CONFINIS Gr. Oak Knoll. August—October.
    S. OCCIDENTALIS Nutt. Arroyo Seco. August—October.

CORETHROGYNE DC.
    C. FILAGINIFOLIA Nutt. Common below 5,000 feet. June—September.

ASTER Linn.
    A. ADSCENDENS Lindl. S. G. Mts. July—September.
    A. EXILIS Ell. Along streets.
    A. FREMONTI Gr. var. PARISHII Gr. Canons S. G. Mts. August—October.

ERIGERON Linn.
    E. CANADENSE Linn. Common below 2,500 feet. June—September.
    E. FOLIOSUM Nutt. Frequent below 3,600 feet. May—August.
    E. PHILADELPHICUM Linn. O. K., A. S. April—July.

CONYZA Linn.
    C. COULTERI Gray. Oak Knoll. July—September.

BACCHARIS Linn.
    B. VIMINEA DC. O. K., A. S., canyons S. G. Mts.

GNAPHALODES Tourn.
    G. CALIFORNICA (F. & M.) Greene. Frequent in dry soil below 3,000 feet. April
        —June.

FILAGO Linn.
    F. CALIFORNICA Nutt. Common in dry soil below 3,000 feet. April—June.

GNAPHALIUM Linn. Everlasting.
    G. CALIFORNICUM DC. Common below 3,000 feet. March—July.
    G. CHILENSE Spreng. Common below 2,000 feet. April—June.
    G. LEUCOCEPHALUM Gr. Frequent between 1,000—3,000 feet. July-September.
    G. MICROCEPHALUM Nutt. Frequent below 2,500 feet. July—September.
    G. PALUSTRE Nutt. Oak Knoll. April—May.
    G. RAMOSISSIMUM Nutt. Oak Knoll. August—October.

AMBROSIA Tourn. Ragweed.
    A. PSILOSTACHYA DC. Com. below 2,000 feet. August—October.

XANTHIUM Tourn.
    X. SPINOSUM Linn. Along streets. June—September.
    X. STRUMARIUM Linn. In waste soil. June—August.

HELIANTHUS Linn. Sunflower.
    H. ANNUUS Linn. Com. below 2,000 feet.

LEPTOSYNE DC.
    L. DOUGLASII DC. A. S. and adjacent hills; foothills S. G. Mts. and adjacent plains.
        June—September.

BIDENS Linn.
B. CHRYSANTHEMOIDES Michx. O. K.; A. S.
B. PILOSA Linn. Along streets. July—October.
MADIA Molina. Tar-weed.
M. DISSITIFLORA (Nutt) T. & G. Along street. July—September.
M. SATIVA Mol. Frequent below 1,800 feet. June—August.
HEMIZONIA DC. Tar-weed.
H. FASCICULATA (DC.) T. & G. var. RAMOSSIMA Gr. Com. Below 1,800 feet.
May—July.
H. PUNGENS Torr. & Gray. Frequent below 1,800 feet. May—August.
H. TENELLA Gray. O. K.; A. S. April—June.
LAYIA Hook. & Arn.
L. GLANDULOSA (Hook.) H. & A. O. K.; A. S. April—June.
L. PLATYGLOSSA (F. & M.) Gray. (Tidy-tips) Com. below 2,000 feet. April—June.
ACHYRACHAENA Schauer
A. MOLLIS Schauer. Near Oak Knoll. March—May.
BAERIA Fisch. & Mey.
B. GRACILIS (DC.) Gr. Com. below 2,000 feet. March—May.
ERIOPHYLLUM Lag.
E. CONFERTIFLORUM (DC.) Gr. Com. in unbroken soil below 3,000 feet. April—
June.
CHÆNACTIS DC. "Pin-cushion."
C. ARTEMISIAEFOLIA Gray. Frequent in dry unbroken soil below 3,000 feet.
June—July.
C. GLABRIUSCULA DC. Frequent on dry hills. May—June.
C. LANOSA DC. Com. below 3,000 feet. April—June.
HELENIUM Linn. Sneeze-weed.
H. PUBERULUM DC. O. K.; A. S., canyons of S. G. Mts. June—October.
ACHILLEA Linn. Yarrow.
A. MILLEFOLIUM Linn. Com. in foothills of S. G. Mts. and adjacent plains. May
August.
ANTHEMIS Linn. Chamomile.
A. COTULA Linn. Com. below 1,500 feet. April—July.
MATRICARIA Linn.
M. DISCOIDEA DC. Com. below 1,500 feet. February—April.
ARTEMISIA Linn. Worm-wood. "Sage-brush."
A. CALIFORNICA Less. (California "Sage-brush.") Com. in dry unbroken soil
below 4,000 feet. May—August.
A. DRACUNCULOIDES Pursh. Frequent below 2,000 feet. May—August.
A. LUDOVICIANA Nutt. Occasional below 2,000 feet. June—August.
A. VULGARIS Linn. var. CALIFORNICA Bess. Common below 2,500 feet. June—
August.
TETRADYMIA DC.
T. COMOSA Gray. Near Devil's Gate. July—August.
LEPIDOSPARTUM Gr.
L. SQUAMATUM Gr. A. S., foothills S. G. Mts. July—September.
SENECIO Linn. Groundsel.
S. CALIFORNICUS DC. Com. below 3,000 feet. March—May.
S. DOUGLASII DC. Hills along A. S., foothills S. G. Mts. July—September.
CARDUUS Tourn. Thistle.
C. CALIFORNICUS (Gr.) Greene. O. K., R. C. May—July.
C. OCCIDENTALIS Nutt. Frequent below 2,000 feet. April—June.
SILYBUM Gaerth. Milk-Thistle.
S. MARIANUM (Linn.) Gaerth. Near Oak Knoll. May—July.
CENTAUREA Linn. Star-Thistle.
C. MELITENSIS Linn. Com. below 2,000 feet. May—August.
PEREZIA Lag.
P. MICROCEPHALA Gr. O. K., A. S., foothills S. G. Mts. June—August.
MICROSERIS Don.
M. LINEARIFOLIA Gr. Com. below 3,500 feet. February—May.

PTILORIA Raf.
  P. CICHORIACEA (Gr.) Greene.   Canyons S. G. Mts.   July—September.
  P. VIRGATA (Benth.) Greene.   Com. below 3,500 feet.   June—October.
RAFINESQUIA Nutt.
  R. CALIFORNICA Nutt.   Hills along A. S.   June—July.
MALACOTHRIX DC.
  M. TENUIFOLIA T. & G.   Com. below 2,500 feet.   June—August.
HYPOCHÆRIS Linn.
  H. RADICATA Linn.   Along streets.   May—July.
CREPIS Linn.
  C. BIENNIS Linn.   Along streets.   April—June.
TARAXACUM Linn.   Dandelion.
  T. TARAXACUM (Linn.) Mac M.   Occasional along streets.   April—October.
LACTUCA Tourn.
  L. SCARIOLA L.   South of Pasadena.   June—August.
SONCHUS Linn.   Sow-Thistle.
  S. ASPER Linn.   Frequent below 1,500 feet.
  S. OLERACEUS Linn.   Common below 1,500 feet.
HIERACIUM Tourn.   Hawkweed.
  H. PARISHII Gr.   Canyons S. G. Mts.   June—August.

SUMMARY OF PLANTS LISTED.

PROTOPHYTA	40
PHYCOPHYTA	50
CARPOPHYTA	350
BRYOPHYTA	53
PTERIDOPHYTA	21
SPERMAPHYTA	542
TOTAL	1,056

INDEX TO FLORA.

www.ingramcontent.com/pod-product-compliance
Lightning Source LLC
Chambersburg PA
CBHW031814090426
42739CB00008B/1266